BOOK OF NEW ZEALAND WILDLIFE

Acknowledgments

Auckland Institute and Museum
Ecology Division, DSIR
Stephen Barnett
Encyclopaedia of New Zealand
Entomology Division, DSIR
Geological Survey
Hauraki Gulf Maritime Park Board
Lands and Survey Department
Brenda May
National Museum, Wellington
New Zealand Automobile Association
New Zealand Forest Service
New Zealand Wildlife Service, for the article 'Work of the Wildlife Service', and for 'The Status of Endangered New Zealand Birds'
New Zealand's Heritage
New Zealand's Nature Heritage
Oceanographic Institute, DSIR
Ornithological Society of New Zealand
The Nature Conservation Council
The Royal Forest and Bird Protection Society
The Royal Forest and Bird Protection Society, for P.H.C. Lucas' 'Role of National Parks and Public Reserves in Nature Conservation'
Royal Zoological Society of South Australia Incorporated

Maps and Diagrams

Automobile Association (Auckland) Inc.
Christine Brown
Christine Halliwell
Jennifer Turley

Design Paper Dart Productions Ltd

Published by Lansdowne Press
26 Customs Street
Auckland, New Zealand

First published 1981
2nd Impression 1982

Typeset by Auckland Typographic Services Ltd.
Printed in Singapore by Toppan Printing Co. (Pte) Limited
38 Liu Fang Road, Jurong, Singapore 22

ISBN 0-86866-061-2

A Guide to the Native and Introduced Animals of New Zealand

C. O'Brien
Chief Cartographer
Automobile Association (Auckland) Inc.
New Zealand

LANSDOWNE PRESS
Auckland • London • New York • Sydney

CONTENTS

ILLUSTRATION ACKNOWLEDGMENTS

Illustrations have been credited by page number. References are made in the order of the columns across the page and then from top to bottom. Acknowledgments have been abbreviated as follows:

AA	Automobile Association
AA (aF)	Automobile Association (after Fleming)
AMA	A.M. Ayling
ATC	Australian Tourist Commission
GB	G.W. Batt
BB	B.D. Bell/Wildlife Service
JB	J. Braggins
CB	C. Brown
PBr	P. Browne
MB	M. Burke
PB	P. Bush/Wildlife Service
JDC	J. D. Campbell
GC	G. R. Chance
PC	P. Coates
LC	L. Cobb
JC	J. W. Cole
AC	A. Cox/Wildlife Service
GD	G. T. Daly
EBD	E. B. Davies
DA	Deerstalkers Association
WD	W. Doak
DOM	Dominion Museum
BE	B. Enting
ED	Entomology Division, DSIR
RF	R. A. Falla
CF	C. A. Fleming
JF	J. E. C. Flux
RF	R. R. Forster
FP	Franz Photographers/Otago Museum
DG	D. Garrick/Wildlife Service
RG	R. Grace
JG	J. Greenwood
JHG	J. H. Green
DH	D. Hadden
CH	C. Halliwell
BH	B. J. Harcourt
PH	P. C. Harper
CLH	C. L. Hopkins
JJ	J. H. Johns/New Zealand Forest Service
RJ	R. R. Julian
E & JK	E. G. & J. I. Kelly
JLK	J. L. Kendrick/Wildlife Service
FK	F. Kinsky/Wildlife Service
PL	P. H. Lucas
TL	T. Lloyd
JM	J. Maslen
RMM	R. M. McDowall
DM	D. Merton/Wildlife Service
KM	K. H. Miers/Wildlife Service
GM	G. J. H. Moon
RM	R. Morris/Wildlife Service
PM	P. Morrison/Wildlife Service
NPS	National Publicity Studios
FS	New Zealand Forest Service
CO	C. O'Brien
LR	L. Richards/Wildlife Service
JR	J. Robb
GR	G. Roberts
CR	C. J. R. Robertson
SP	Shell Oil (N.Z.) Ltd
DS	D. Scott
RS	R. Smith
DSED	D. C. Smith/Entomology Division DSIR
CK	C. Smuts-Kennedy/Wildlife Service
MS	M. F. Soper
FT	F. Thompson
MT & BM	M. M. Trotter & B. McCulloch
TL	Turnbull Library, Northwood Collection
TU	T. R. Ulyatt/Geological Survey
UNP	Urewera National Park Board
CV	C. R. Veitch/Wildlife Service
DW	D. Whyte
WS	Wildlife Service
AW	A. Wright/Wildlife Service

Title page: E & JK
Contents verso: AW
Opposite: E & JK
Pages: **8** TU: **9** PBr: **10** AA: **11** JC: **12** AA (aF): **13** AA:
14 RM: **15** TL: **16** GM: **17** AA/FP: **18** MT & BM/SP:
19 JDC/AA: **21** AA: **22** E & JK: **23** AA: **24** CH/AA:
25 E & JK/BE: **26** WD/WD: **27** CH/GB: **28** GR/AMA/RF:
29 E & JK/GM: **30** AA/E & JK: **31** AA/GM/E & JK/E & JK:
32 AA/RF/E & JK: **33** JR/JR: **34** E & JK/CO/AA:
35 CV/CO/WD: **36** E & JK: **37** FS: **38** FS/E & JK:
39 RM/JLK: **40** FS/AA/FS: **41** E & JK/JR/E & JK:
42 GB/ED/E & JK: **43** E & JK/ED/CO:
44 CB/JM: **45** E & JK/E & JK: **46** WS/MS/E & JK: **47** DH: **48** JK/PM/MS:
49 JLK/MS: **50** JG: **52** AA/AA/GR:
53 CK/JLK: **54** JLK/AA: **55** GR/CH: **56** AA/FS/DA:
57 FS: **58** E & JK/E & JK/EBD: **59** GR/CO
60 DG/JR/JR/CV: **61** E & JK/CV/JJ: **62** CO/FT:
63 RF/RF: **64** DOM/RF: **65** AA/DSED/DSED: **66** E & JK/E & JK/EBD: **67** E & JK/RS/GB/JHG: **68** CB: **69** JLK/AA:
70 RM/GR: **71** GM/DM: **72** E & JK/PM:
73 E & JK/MS: **74** E & JK/E & JK/E & JK/E & JK:
75 GM/DM: **76** MS: **77** RS: **78** UNP/FS:
79 CO/MS: **80** E & JK: **82** AA/AA/FS:
83 E & JK/RM: **84** JF/GR/GR: **85** JR/CF: **86** E & JK/ED/ED/AA:
87 CB/RF: **88** GR/GC: **89** E & JK: **90** AA/GM/RM:
91 CH/RM: **92** MB/CO: **93** LC/CV: **94** E & JK:
96 DS/CB/RM: **97** RMM/RMM/RMM: **98** RMM/RMM:
99 RMM/RMM/RMM: **100** GR/AA/RMM: **101** CO: **102** RMM/CLH:
103 E & JK/RJ/E & JK: **104** CO/JLK:
105 E & JK/E & JK:
106 RM/GC: **107** RM/FS/GM: **108** MS/CK:
109 CO/LR: **110** CV: **112** AA/JR/GM:
113 CH/CV/RG: **114** CK/E & JK:
115 PH/PM **116** RM/JLK:
117 DW/CV/GM: **118** ED/E & JK:
119 GM/ATC/FS/E & JK:
120 JF/JR: **121** GM/E & JK:
122 JR/DG: **123** CR/ED/NPS:
124 JLK/AA:
125 MS/ED/PC:
126 GR: **127** BB/CV: **128** WS/AC:
129 AA/CV/FK/MS:
130 MS: **131** JJ/BH: **132** BH/E & JK: **133** FS/AW/TL:
135 JB/FS: **136** WS/DM/JLK/JLK: **137** GM/JLK/RM/RM/CV:
138 JLK/JLK/CV/CV: **139** CV/JLK/KM/AC: **140** JLK/RM/AC/AW/CV/RM:
141 JR/JR/DG/JR/CV: **142** RM: **143** PB/WS:
144 BH: **145** AW/WS: **146** PL/PL: **147** PL: **148** PL/PL:
149 PL/PL: **150** PL/PM: **151** WS:
152 RM/WS **153** GD:
154 CO/ED/MS

BEGINNINGS

To understand New Zealand's present fauna one must delve for the necessary background into the realms of geological science.

New Zealand is not ancient by comparison with the vast antiquity credited to the continental masses; nevertheless, there is irrefutable evidence in our rocks that our region reaches back at least 600 million years, that is, to the Cambrian, the universal base line marking the first appearance in geologic strata of well preserved and varied animal remains that constitute a faunal assemblage.

In visualising New Zealand's past history, one is prone to think in terms of this country's present size, shape, and approximate topography, but all three are very recent and very transient features. We must, therefore, project our minds back in time and become accustomed to the idea of a fluctuating land mass sited around a place on the globe marked by the present location of the country. We must envisage a land of continental dimensions at one stage, a mere scattering of small islands, an archipelago, at others – a land not entirely coincident with its present location but sometimes extending greatly both westward and northward. The actual site of New Zealand as we now know it was a trough in the ocean bed which steadily accumulated sediments that were slowly built up, and in time became raised to the surface. Their contained fossil remains now give us the clues to our past history.

Left: *Banded gneiss on the Westland Coast – New Zealand's oldest rocks.*

Volcanic activity continues to play a significant part in the formation of an ever-changing landscape.

Climate is another consideration, for the fossil faunas reveal fluctuations ranging from near tropical to conditions far colder than those experienced today. For instance, *Conus* shells, characteristic of warm seas, lived here in early Tertiary times but are absent from our seas of today; coconut palms flourished in North Auckland in comparatively recent times but disappeared during the world-wide ice ages, when New Zealand, although it escaped many of the rigours of northern lands, did see a great advance of glaciation, particularly in the South Island. Despite all these changes, New Zealand has existed as a more or less separate entity for a very long time, at least as far back as the early Tertiary, 50 million years or more ago, and during that time it has evidently had no direct connection with the Australian continent or with other southern lands. Nevertheless, there is evidence of the arrival of new colonising elements from time to time but not of necessity those which required a continuous land connection for the purpose.

One of the main colonising agencies at the present time is the East Australian warm water current which originates in the New Caledonian area, sweeps down the New South Wales coast, and then across the Tasman in a great arc, bringing with it not only adult long-ranging swimmers that are induced to travel southward of their normal haunts, but also larval forms of marine organisms, many of which ultimately become acclimatised.

Long isolation from Australia has prevented land mammals from reaching New Zealand; the only endemic land mammals are two small species of bats. On the other hand, the greater part of our present land bird fauna is largely of

Australian origin; the sea is not a formidable barrier to creatures with well-developed powers of flight. The fact remains that there is no geological evidence to suggest a direct trans-Tasman land connection with Australia at any time.

It is evident, however, that within comparatively recent geological times, transient land extensions to the north, to the vicinity of Melanesia, allowed the influx to northern New Zealand of fauna and flora of New Caledonian origin; not necessarily a continuous land connection, but rather a rafting effect of a series of give and take land movements.

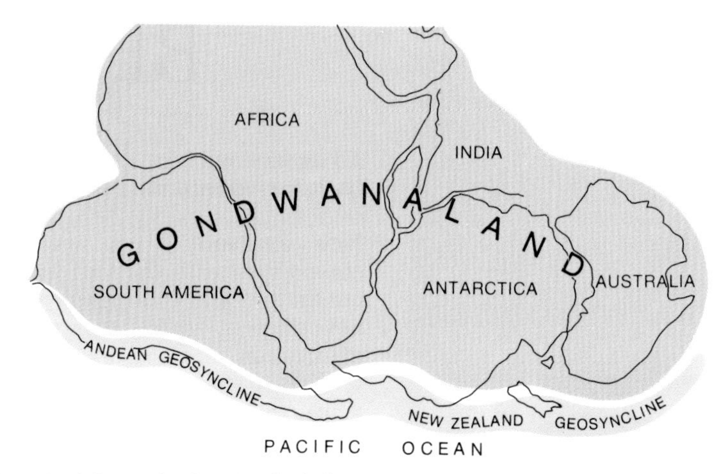

Gondwanaland during the Permian Period.

In the flora we have the striking examples of the kauri tree, mainly of Melanesian distribution, and a 'bottle-brush' flowering shrub found only in New Caledonia and islands off the North Auckland coast. In animals, there is the influx to the North Auckland Peninsula of *Placostylus,* a genus of large land snails, otherwise restricted to the islands of the now largely submerged Melanesian Plateau.

Archaic elements in our land fauna are the reptile tuatara, the moas and allied genera, the kiwis and the primitive caterpillar-like arthropod, peripatus. The moa belongs to that group of large, flightless Southern Hemisphere birds which includes the Australian emu, the cassowary of New Guinea, the ostrich of South Africa, the rhea of South America and the extinct *Aepyornis* of Madagascar. Some interconnection of southern lands in the distant past is required to account for the present wide dispersal of a group of ponderous birds ranging in height from one to 3.5 m. It is hard to visualise how this dispersal was achieved, but the most widely accepted theory is that of the one time existence of a single southern continent – Gondwanaland.

The Drifting Continent

The theory of continental drift is an old one. In 1620 Francis Bacon in England speculated that America had been joined to Europe and Africa. Last century the great Austrian geologist, Eduard Suess, found such close similarity in the geology of southern lands, including India, that he decided they were once linked as a single continent which he called Gondwanaland after an Indian province. In 1910 Alfred Wegener, a German, suggested that all the continents had once been joined as a single land mass, Pangaea, which had broken into parts, which then drifted to their present positions. Such a hypothesis seemed so unlikely and so contrary to then current geophysical theory that it remained unpopular for most of the century, although supported by the geological researches of such workers as A. L. du Toit of South Africa, S. W. Carey of Australia, Lester C. King of New Zealand and Vening Meinesz of the Netherlands. Further support for Wegener's theory came from evidence of horizontal movements over hundreds of kilometres of the earth's crust along fracture faults such as the Alpine Fault in New Zealand and others in Scotland and California.

Vindication of a Theory

During the 1960s a great change in thinking was brought about by evidence arising from research in several fields. The topography of ocean depths had been mapped in detail by echo sounding; and accurate location of earthquake centres had led to better definition of earthquake zones. Measuring the radioactivity of minerals also gave the age of belts representing ancient, deformed mountain roots, while studies of rock magnetism led to several crucial conclusions.

When rocks are formed, the iron in them, like a 'frozen' compass needle, records the direction of the earth's magnetic field at the time of their origin. The precise measurement of magnetic direction in ancient rocks shows that the magnetic poles have wandered throughout geological time. In Britain, Blackett and Runcorn found that the paths of such polar wandering change from continent to continent but agree more towards the present time, thus suggesting that the continents have shifted over the surface of the earth to their present positions.

But the most dramatic evidence came from oceanographic exploration. The amount of sediment – sand, mud and lime deposits – on ocean floors seemed too scanty and too young to be the produce of earth's total history and all the oceanic volcanoes are also young; it therefore appeared that ocean floors must have been constantly rejuvenated in some way.

It was found that the ocean contained central ridges characterised by high emission of heat and earthquake activity and crossed by fault fractures. In 1960 H.H. Hess and R.S. Dietz independently suggested that rising convection currents at these mid-ocean rises produced new ocean bottom material which spread outwards, sweeping sediment toward the continents, and thus preventing accumulation of ancient rocks. It was a bold theory – soon to be confirmed by marine geophysics.

By extending the Hess-Dietz hypothesis it was suggested and subsequently proved by computer techniques that molten rock, rising at the mid-ocean ridges, formed a continually spreading ocean floor. Since 1966 this concept of sea-floor spreading has led to revolutionary changes in ideas of earth structure and to the acceptance of the theory of continental drift. According to the theory the continental slabs can be carried like icebergs on currents established by the spreading of the sea floor. The directions of these currents are quite compatible with the main movements necessary to separate the jigsaw of an assembled Gondwanaland into its various parts. The fact that many oceans are bounded by deep, elongate trenches also supports the theory, for these are now seen by many scientists as the 'sinks' into which the ocean crust descends below the continental slabs.

Gondwanaland, during its long history, supported a distinctive vegetation of early fern-like plants and conifers. Of these, the best known is *Glossopteris,* a seed fern that has been found in characteristic sequences of Gondwana rocks in South America, South Africa, Antarctica, India and Australia. Glacial deposits indicate that the Gondwanaland continent underwent a succession of glaciations and was thus close to the South Pole, a conclusion supported by the magnetic evidence from its ancient rocks. The continent also supported a rich animal life – primitive freshwater fishes, amphibians and reptiles, which required continuous land for their dispersal.

The place of New Zealand in the history of the dissolution of Gondwanaland is still uncertain. During the Permian, Triassic and Jurassic Periods – roughly 300 million to 135 million years ago – the area now called New Zealand apparently lay near the margins of the Gondwanaland continent. It consisted of a vast, linear, trough-like feature, called by geologists the New Zealand geosyncline, which extended from New Caledonia to the Chatham Islands, and which received abundant marine sands and muds and volcanic ash from the adjacent lands and from volcanic island arcs. The geosyncline is now bent in an S-shaped curve, but was then probably much simpler in form.

The characteristic Permian plants of continental Gondwanaland were sometimes washed out to these marine deposits, most of which are now hardened and contorted to form the schists and greywackes of the main mountain ranges. Only recently has *Glossopteris* been found in New Zealand and so far no freshwater or land animals have been found here as fossils, though they are abundant in parts of Antarctica.

Later, in the Triassic and Jurassic, plant remains were preserved in New Zealand, including probable ancestors of the conifers – relations of kauri and podocarps – that have remained characteristic of southern lands. Although there is no fossil evidence, it seems quite probable that ancestors of the tuatara, *Sphenodon punctatus,* and the native frog, *Leiopelma,* and other primitive animals, peripatus, as well as ancestors of the kauri, rimu, totara and other pines, reached New Zealand before the dissolution of Gondwanaland.

New Zealand Becomes Isolated

Disruption of the Gondwana continent, according to one interpretation of the data on sea-floor spreading, began with the separation of South America from Africa in the Jurassic, about 150 million years ago. In the early Cretaceous, 135 million years ago, Africa moved away from India, Australia, New Zealand and Antarctica, and later, in the Cretaceous, New Zealand separated from Antarctica. The New Zealand geosyncline which had formed the marginal sink zone for the ancient south-west Pacific suffered a period of violent upheaval in the early Cretaceous, probably during a reorganisation of the global sea floor spreading pattern.

Flowering plants, angiosperms, arose and began to disperse throughout the world in the early Cretaceous. Before New Zealand was completely isolated from other southern lands, some of the angiosperms with restricted ability to disperse had reached these shores. The southern beeches, *Nothofagus,* characteristic of South Island forests, also occur in Australia, New Guinea, New Caledonia and South America, although botanists believe they are unable to disperse across the sea. Their remains, leaf impressions and characteristic pollen

The volcanoes of Tongariro National Park looking south and showing the alignment of the vents.

grains, have been recognised as fossils back to late Cretaceous in New Zealand and they also occur in Antarctica, where the climate at sea level was warm enough to support forests in the Tertiary, between 70 million and 1 million years ago. Apparently, then, some links between Gondwanaland's fragments lasted into the late Cretaceous, and if so they would have allowed the ancestors of the flightless moas and kiwis, and perhaps other distinctive animals to reach New Zealand.

The isolation of New Zealand as an island continent, separated by ocean barriers from all other lands, must have been complete before the spread of snakes and mammals which were evolving in the Cretaceous. There is no evidence that they were ever in New Zealand, although early representatives of the mammals reached Australia to produce the monotremes – the platypus and echidna – and the marsupials.

The early Cretaceous upheaval known as the Rangitata Orogeny, increased the land area of New Zealand, but the sea gradually spread over a land that had been worn down into a plain across the roots of the early Cretaceous mountains. Structurally, and therefore geographically, Tertiary New Zealand became a good deal different from Mesozoic New Zealand. Instead of a broad trough some 200 to 400 kilometres wide and thousands of kilometres long, the land moved up and down as a series of narrow, short, branching folds. The welts, which tended to be submarine ridges or land, were small, so Tertiary New Zealand could be described as a changing archipelago. Changes in geography were continuous. Troughs sank rapidly but filled with sediment as they subsided, therefore seldom reached abyssal depths. Welts rose between the troughs, but owing to constant erosion were seldom mountainous. Throughout the 60 million years of Tertiary time, colonising plants and animals had to cross the ocean to reach the New Zealand islands.

A kind of writhing of part of the mobile Pacific margin seems to have occurred in the late Tertiary, culminating in the Kaikoura Orogeny in the early Pleistocene, between 1 million and 2 million years ago, when uplift of the main range in a period of cooling world climate brought glaciers to New Zealand for the first time.

The older elements of New Zealand's plants and animals date back to Gondwanaland. They include several plant species, the archaic tuatara, peripatus and moas already mentioned. Later, throughout the 100 million years since the Lower Cretaceous, these ancient New Zealanders were joined by colonists from Australia, the nearest land, and from the tropical Malayo – Pacific to the north. As world climate cooled in the Pliocene, about 13 million years ago, the cool, west wind drift brought a new wave of southern colonists and other immigrants of northern affinity, perhaps at a time when birds were developing the habit of trans-equatorial migration that enabled them to exploit the seasonal luxuriance of the northern tundra.

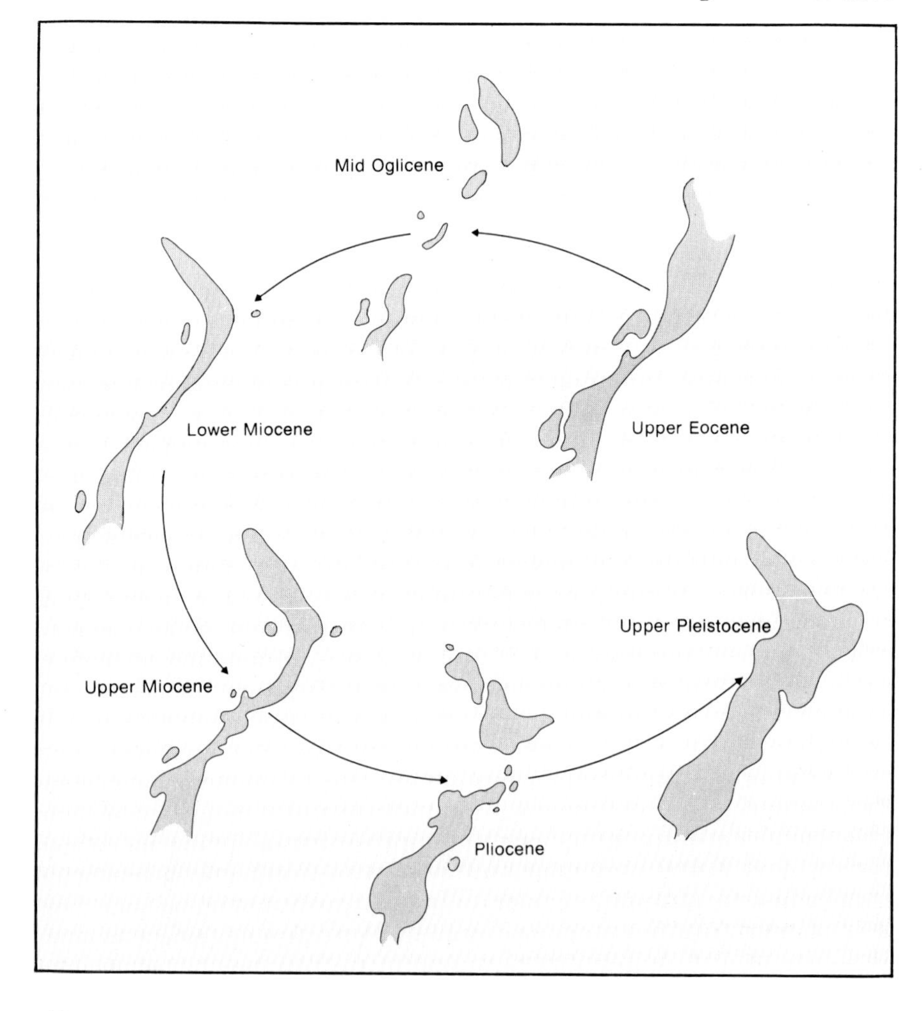

The Coming of Man

So far as is known, every animal that landed in New Zealand came by sea or by air. The fossil record shows that for long periods little or nothing of the land showed above the waterline. Old rocks – and some of New Zealand's go back before the Cambrian, when 500 million year old signs of life first appear – are by no means the same thing as old land. Fossils are mostly found in marine deposits, among the sand and silt and mud of ancient sea beds; sometimes in old lake beds; and the fact that they are still there at all is pure luck for the geologist. They have had durable, hard shells or other parts that have lasted and they happen not to have been destroyed by volcanic eruption or other natural accidents. Sketchy as it is, this fossil record tells much about marine invertebrate life, of the tiny shelled creatures, sea urchins, starfishes, molluscs and so on, that flourished and decayed over millions of years. Vertebrate fossils, those of animals with bones, also survived. There were swimming reptiles in the New

Opposite: *Biological events in relation to the Geological time scale.* Left: *The New Zealand coastline throughout the Caenozoic era.*

SCALE: C, M, P, PRE-CAMBRIAN (millions of years: 1000, 2000, 3000, 4000, 5000, 6000)

ERA	PERIOD	EPOCH	THE CHANGING EARTH	BIOLOGICAL TIME SCALE	DEVELOPMENTS IN N.Z.
CAENOZOIC	QUATERNARY	**Holocene** *Started 20,000 years ago*	Man learns to domesticate animals and cultivate plants.		Fossil tuataras and avifauna.
		Pleistocene *Started 2 million years ago*	Stone-age man originates in Africa. Ice ages alternate with warmer periods. Marine life much as it is today.	*Man*	Fossil land insects and molluscs.
	TERTIARY	**Pliocene** *Started 10 million years ago*	Mammal species decline with exception of man-like apes. Continents and oceans begin to take on their present form.		Pseudodontornis of Motunau; Galaxias.
		Miocene *Started 25 million years ago*	Further development of deciduous trees. Much volcanic activity. Bony fish continue to increase in variety and size.		Renewal of mountain building. Fossil coconuts.
		Oligocene *Started 40 million years ago*	Herbivorous mammals flourish in grassland areas. Bats abundant. Insects become of modern type.		Fossil squid, octopuses, tusk shells. Fossil crabs and other crustacea. First whales.
		Eocene *Started 55 million years ago*	Disappearance of the dinosaurs and spread of smaller mammals. Flora takes on modern appearance, ferns and conifers plentiful.		Ancestors of present-day trees.
		Paleocene *Started 65 million years ago*	Considerable alteration between land and sea. Some mammals become semi-aquatic and fishes become similar to those of today.		
MESOZOIC	**Cretaceous** *Started 135 million years ago*		Vegetation flourishes. Heyday of the dinosaurs, giant reptiles and turtles. Birds continue to evolve.		Giant marine reptiles. First sharks. Break-up of Gondwanaland. First flowering plants. Opening of Tasman Sea.
	Jurassic *Started 190 million years ago*		Mammals remain small and primitive. Reptiles increase. First bird evolves feathers from scales.	*Birds*	Appearance of Nothofagus and Proteaceae. Ancestors of tuataras, moas, kiwis and podocarp. NZ above water – first mountain-building stage.
	Triassic *Started 225 million years ago*		Hot and dry climate prevails. The first mammals evolve from the reptiles. Appearance of king dinosaurs. Insects abundant.	*Mammals*	Fossil traces of worm species. Ferns and fern allies dominate Gondwanaland. Ichthyosaurs – oldest vertebrate fossil. Dinosaurs roam most of the Earth.
PALAEOZOIC	**Permian** *Started 275 million years ago*		Ice-age continues in southern hemisphere. End of domination by marine creatures. Land animals and plants increase.		Fossil solitary corals. Glossopteris fossil. First Ammonites.
	Carboniferous *Started 350 million years ago*		Chief coal forming period. Fishes common, amphibians increase. Luxuriant vegetation includes giant evergreens.	*Reptiles*	Gondwanaland at South Pole.
	Devonian *Started 400 million years ago*		Extensive mountain building and volcanic activity. Earth begins to look green. Rapid evolution of vertebrates and wingless insects.	*Amphibians*, *Insects*	Disappearance of Trilobites. First corals. Oldest bivalve fossils.
	Silurian *Started 440 million years ago*		Plants first adapt to life on land. New species of marine vertebrates including huge heavily armoured sea-scorpions.	*Fish*, *Land plants*	
	Ordovician *Started 500 million years ago*		All life still restricted to water, plant life still confined to seaweeds. First vertebrates appear: the ostracoderms. Shallow seas retreat.	*Vertebrates*	Fossil Echinoderms. Graptolite fossils.
	Cambrian *Started 600 million years ago*		Shallow seas cover much of the Earth. Evolution of invertebrates: jellyfish, worms, sponges and trilobites.		Oldest sea-snail fossils. Oldest NZ fossils. Oldest NZ rocks. Formation of Gondwanaland.
Pre-Cambrian			Seaweeds only form of vegetation. Primitive soft-bodied invertebrates originate in warm seas.	*Marine invertebrates*, *Seaweed*	Earth's first forms of life. Oldest rocks of continental land mass.

Without man, most of New Zealand would still be a jumble of primeval forest with animal and plant life evenly balanced according to its environment.

Zealand region in the Mesozoic, penguins in the Oligocene, also whales.

Yet although over the centuries the land was full of birds, they have left no fossil traces. Some might explain this dearth by the proclivity of the bush to swallow up animal remains, digested by its wet, acid leaf-litter and fallen, decaying logs. There is even no certainty that New Zealand's two native bats are the only mammals the country had. And there can only be speculation about ancestral birds, and the contemporaries of the original tuatara. The tuatara established itself in the area during the Age of Reptiles, perhaps coming by land across a vanished continent, but fossil tuataras have not been found here, only related ones, extinct long ago.

It is normal for large land masses to support mammalian life, as well as other vertebrate groups. Australia, of course, has its unique mammalian population. But until man came, birds were the dominant form of life in New Zealand. In their day the land had looked more like Captain Cook's chart than it does now, with a real Banks Island and no Foveaux Strait, or only a shallow one, and perhaps the climate was warming a little as the great glaciers receded. There must, at any rate, have been open grazing country where vast flocks of flightless birds stomped about, protein for the new arrivals.

So moa country became man country. The kuri(dog) and the kiore(rat) that the newcomers imported from the Pacific seem not to have made much difference to the landscape. But further human migrations from the Pacific meant less room, and less food to go round, and when the moas became scarce, other birds had to be snared or knocked down, fish had to be caught, seals hunted and shellfish fossicked at low tide.

Archaeologists may never discover exactly when and where the first settlers to reach New Zealand made their landfall, for the faint traces of those few earliest arrivals may long since have vanished. Careful investigations in many parts of the country however, have shown that by A.D.1000 or 1100 a substantial population was already established. Village and camp sites dating back to this time have been found in many places around the coasts from Auckland to Otago, and people had already explored inland to the shores of Lake Taupo and the inhospitable regions of Central Otago.

As a result of careful excavation of their settlements a good deal is known about these people, whom archaeologists call moa-hunters or New Zealand East Polynesians. They came from small tropical islands of East Polynesia, such as Tahiti, the Marquesas and the Cook Islands, the area always regarded as the traditional Hawaiiki of the Maori. They looked very like the Maoris whom Cook saw in 1769, but their fashions in clothes, ornaments and tools were more like those of the islands from which they had come, than those which Cook and his men obtained from Maoris.

But it was not until after the European arrival that major change came to the land without mammals. It has been noted that many of the settlers' diaries dwell on the loneliness of a landscape that, initially, contained no hayricks, church towers, wagons, horses, cattle and sheep – only bush, and more bush. With this outlook they were not likely to arrive at a conclusion that a land virtually without mammals was the wrong place to bring mammals. Were there not wild animals in every northern forest? So inevitably the land that, left to itself, might have maintained a vegetation-birdlife balance was never the same again. Fire and axe, deer, opossums from Australia, wild cats, rabbits and the rest all took their toll of the vegetation. But the vegetation contained birds, and also held together many a steep mountain slope. Now when a rain sodden mass was carried away, it was slow to regenerate.

Yet there were other aspects of a unique ecology that were tragically vulnerable to change. Rats, coming ashore from tied-up muttonbirders' boats, quickly multiplied and eventually eliminated the bush snipe, saddleback and wren from the South Cape Islands, repeating the elimination on the mainland of Stewart Island. Similarly, the Stephens Island wren was no match for the lighthouse keeper's cat. With no predators to spring at them, so many New Zealand birds had previously scuffled about among the ferns and mosses of the bush floor, as wekas and kiwis still do. And they were unable to adapt to the changed circumstances.

Today, there are few traces left of the old land without mammals. The New Zealand Forest Service has brought back something of it in microcosm, with its enclosure plots of bush fenced off from pests; the small islands in Paterson Inlet, Stewart Island, are natural enclosure plots, so long as they are kept free of swimming deer. As much as possible the Forest Service and other government agencies like the National Parks Board work towards the careful preservation of the remainder of the land's ancient cover.

It may well be that part of the settlers' dilemma was their feeling of kinship with fellow mammals. A man looks a dog between the eyes and the dog looks back – seeing a man no doubt in shades of grey, for smell replaces colour in its world – but its eyes hold an expression the man does not see in the beady eyes of a bird, quick and bright as it is; not in the eye of the lizard, shark or octopus, advanced as these are in the matter of seeing. Europeans had been brought up among other mammals. They felt the need to have them in their new landscape, to provide food and clothing, sport and, perhaps most of all, companionship.

Introduced Mammals

Once the mainland of New Zealand was settled, a brisk trade sprang up in introduced birds and mammals from Europe; numerous local acclimatisation societies organised the shipments, looked after the animals on arrival and distributed them around the country. These societies were well organised and their aim to enrich the land 'with the choicest products of the animal kingdom' must have seemed attractive and praiseworthy to the early settlers, dumped so far from home in a land with few native mammals. To their credit, most of the animals they introduced did become established, though nowadays with hindsight New Zealanders may look askance at their successes.

Red deer and hares were introduced in 1851, and the first wild rabbits that followed in 1864 were immediately successful. Fallow deer were also introduced in 1864 and sambar (originally from Sri Lanka) in 1875-76; the first hedgehogs were released in 1870. Ferrets, used for hunting rabbits, were brought in three years after the rabbits themselves, an obvious indication that rabbits were thriving – indeed they were already recognised as a major pest by the early 1880s, when stoats and weasels and more ferrets were hastily imported to combat their spread. The Department of Agriculture of the time organised an intensive programme for breeding ferrets and released many thousands.

Meanwhile, about 1876, Governor Sir George Grey, perhaps impressed by the success of the

Muscle-power provided by bullock teams was crucial to developing communications and transforming a primeval wilderness into an agricultural economy.

opossum on the mainland, had introduced five wallabies from Australia to Kawau Island in the Hauraki Gulf, and a sixth species was liberated in the South Island a few years later. Wallabies were not the only mammals that Sir George released on Kawau Island; he included a solitary zebra, which of course died without progeny. One might question whether the menagerie of animals assembled on Kawau Island should really qualify for inclusion among the wild mammals of New Zealand; but the fact is that at least four of the five wallaby species released on the island survive there today.

In the first decade of this century no less than seven more large game animals were established to supplement the red deer, fallow and sambar already in New Zealand. These additions were Japanese or sika deer from Asia, rusa from New New Caledonia, wapiti, moose and Virginia deer from North America, chamois from Austria and Himalayan thar. They met with varying success. The Japanese deer is now widely distributed in the centre of the North Island; rusa are confined to a small herd near Rotorua; wapiti to Fiordland (where they interbreed with red deer) and a few moose may still survive at Dusky Sound in Fiordland; Virginia deer are thriving on Stewart Island and there is a small herd near Lake Wakatipu in the South Island; chamois have spread from Mount Cook almost the full length of the Southern Alps from Nelson to Fiordland, while thar are still restricted to the Alps between the Waimakariri River and Lake Wanaka.

No new species of exotic mammals has been introduced since 1910. Yet it is unrealistic to suppose that, in the long term, legislation alone can do any more than slow down the colonisation of New Zealand by mammals from overseas. Small, inconspicuous mammals are perhaps the most likely species to take advantage of the increasingly rapid and frequent sea and air traffic, and it is these small species that are conspicuously lacking from the New Zealand fauna.

Each species of animal is a specialist in a particular way of life: rabbits eat short herbage, mice eat small seeds, deer eat longer herbage and shrubs, opossums eat leaves of trees and shrubs, and stoats and cats devour other animals usually smaller than themselves. The number of different kinds of animal that can live in the one place does not depend just on the population, but also on the variety of habitats available with sufficient food and shelter.

Sir George Grey liberated the first wallabies on Kawau Island where four species are now well-established.

If one compares the present list of New Zealand mammals with that of other comparable but less isolated groups of islands – for example the British Isles or Japan – it becomes clear that New Zealand has many large mammals such as the various species of deer, yet few small mammals such as bats, mice, voles and shrews. There can never have been much incentive to introduce such tiny creatures, and only the rats and mice that could live 'wild' on board ship ever arrived in the country.

Deliberately or otherwise, man has allowed many animals to reach New Zealand. He has also provided a much wider range of habitats than before by converting forest to pasture, growing crops, impounding lakes, and even by building towns and cities. Virtually all this took only a couple of hundred years, whereas in countries long settled by man it happened gradually over thousands. It is hardly surprising that the sudden arrival of the European, with such an array of introduced mammals, has profoundly altered New Zealand's native fauna and flora.

Native Casualties

In the early 1800s approximately 66 per cent of New Zealand was forested but now only about 22 per cent of the original forest cover remains, and most of that has been highly modified by browsing, grazing or logging. Initially the land was cleared for farming and forestry but more recently the emphasis has changed to clearance for conversion to exotics. Proportionately more of the lowland podocarp forests have been cleared and this has been serious because these have a more diverse avifauna than the beech forests or forests at higher altitudes.

The loss of native forests has not been the only problem. Introduced browsing animals such as goats, deer, and Australian opossums have severely modified all but a small proportion of the remaining forests

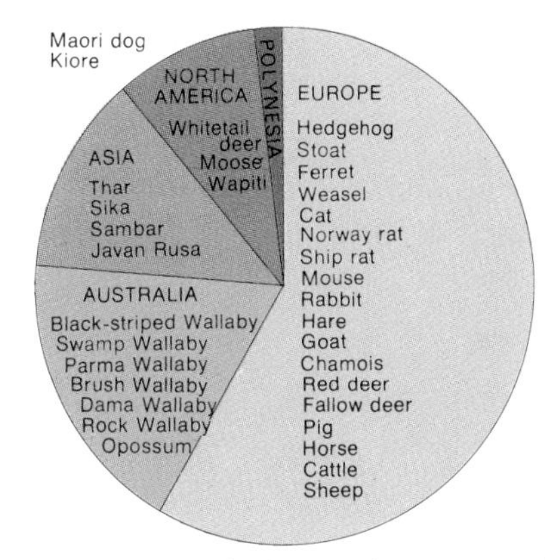

Introduced animals showing their regions of origin.

and have had a detrimental effect on the alpine and subalpine grasslands (both having evolved in the absence of mammalian herbivores).

It is evident also that the introduction and spread of the ship rat, the Norway rat, stoats, ferrets, weasels, and cats have had an important impact. A survey of the predatory behaviour of the three species of rats has indicated that the ship rat concentrates on eggs and nestlings of tree-nesting forest birds, the Norway rat prefers the eggs and nestlings of ground-nesting birds and the kiore the small eggs and nestlings of ground and forest-nesting birds.

The spread of the ship rat was probably also responsible for a major decrease in the numbers of several species of native forest birds. In the North Island the spread is believed to have occurred after 1860 and this coincided closely with a decline of the bellbird, robin, stitchbird, saddleback and North Island thrush. Similarly, in the South Island the spread of ship rats which occurred after 1890 coincided with a decline of the saddleback, robin, bellbird, kokako and red and yellow-crowned parakeets; but the possible effects at this time are difficult to separate from those attributed to mustelids which were also proliferating at the same time. The arrival and spread of the Norway rat in the eighteenth century predates the major period of decline of the avifauna.

A recent example of the severe impact that the ship rat has had on New Zealand's avifauna is graphically illustrated following their arrival in 1964 on three islands west of Stewart Island (Big South

How the moa Dinornis robustus *probably looked – from a reconstruction on display in the Otago Museum.*

Cape (911 ha), Solomon (30 ha) and Pukaweka). Within three years of coming ashore from fishing boats the rats reached plague proportions and four species of indigenous land birds were eliminated – Stead's wren, the Stewart Island snipe, the Stewart Island robin, and the Stewart Island fernbird. Four species were greatly reduced – the red-crowned and yellow-crowned parakeets, the South Island saddleback, and the bellbird. Within another five years the saddleback had also disappeared. In addition to the reduction in avifauna, a rare species of bat almost disappeared and the invertebrate fauna became depleted. The introduction of diseases to populations which had no immunity must also be taken into account.

Following the decline of some bird species on the mainland of New Zealand last century there was a resurgence of some others such as the bellbird, the red- and yellow-crowned parakeets, brown creeper, kiwi, whiteheads and perhaps tits. This resurgence could be due to the species adjusting to the changed conditions or to a stabilising of the predator populations following the initial irruption.

The preponderance of birds in the primitive New Zealand ecology and their developing unique forms was due to the island chain breaking off from the main land mass at a time (70 million years ago), pre-dating mammal migration to the region. This allowed a situation in which bird life was free from predation. Many forms became flightless, or in the main, ground-dwelling, with reduced wing-bones and lengthened legs.

Probably best known amongst these is the moa which comprised a group of large flightless species and included the tallest bird ever recorded. They belong to the Ratite order of birds, (so named because of the absence of a keeled sternum, present in all flying birds and to which is attached the strong wing muscles – moas could not only not fly, they did not even have any wing-bones). Other Ratites include the kiwis, ostriches, emus and rheas.

The range of moas in pre-Tasman Sea days seems to have extended as far as northeast Australia, based on the finding of a moa femur in Queensland, so that moas possibly originated in the Australia-New Zealand region, to be later isolated in the latter. They apparently had no enemies in New Zealand and evolution proceeded at a steady rate, resulting in a graded series of species with a marked break into two families. The 20 or so species ranged from *Dinornis maximus* at 3-4 m tall – taller than any living animal except the giraffe and African elephant, and possessing leg bones more massive that those of a draught horse – to smaller types of about 1 m. The larger ones were ponderous types suited to grazing the eastern South Island, and eastern and south-eastern North Island grasslands. The moa species on the western side of the dividing ranges of both main islands were similar in being smaller, more mobile birds better suited to forest life.

The oldest moa bones date back 2-7 million years and are evidence that moas survived the alternating ice ages and warm interglacial periods of the Tertiary Era. From the great numbers of bone deposits that have been found, the moas must have roamed in their thousands at the close of the last Ice Age 20,000 years ago.

Indications are that moas were dying out by the time the first Polynesian settlers arrived in New Zealand, although the reasons for this decline are not known. Evidence from the middens at moa-hunter camps suggests that all genera except *Euryapteryx* and one other were extinct or greatly reduced by the time the first humans arrived. And there is no record of any species of moa ever having been seen by a European, although one of the smaller forest species could well have survived through to the last 200 years.

In addition to some 24 species of moas considered extinct at the time of European contact, there were a number of other birds, now known from fossil and subfossil remains. The list comprises penguins (including the giant penguin standing approximately 1.7 m and estimated to have weighed up to 100 kg), hawks, eagles, swans, geese, rails, ducks, a giant pelican, a snipe, a crow and a large prehistoric bird with a toothed jaw, *Pseudodontornis stirtoni,* which is thought

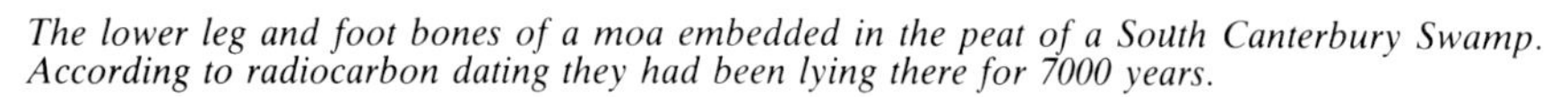

The lower leg and foot bones of a moa embedded in the peat of a South Canterbury Swamp. According to radiocarbon dating they had been lying there for 7000 years.

An artist's impression of Pseudodontornis, ('bird with false teeth'), huge fish-eating bird with a wingspread of about 4.5 m, thought to have lived some 5 million years ago.

Broken River, Castle Hill Basin, a limestone gorge below the mouth of Coal Creek where fossils have been collected, the first about 1866 by naturalist J. D. Enys, from the tuffs of the Thomas Formation (left foreground).

to have looked like something between a pelican and a shag.

Even in the last hundred years several animals, particularly birds, have either succumbed to extinction or remain as species in grave danger of extinction. The huia, piopio and South Island kokako are only three that have vanished from the forests in recent times. In fact New Zealand has an unenviable record: casualties and endangered species are disproportionately high, and New Zealanders must become more conservation conscious if those species are to survive.

Classification of Living Things

For hundreds of years Man has attempted to classify and categorise the teeming varieties of plants and animals as a necessary step in increasing his understanding of the world.

The science of classifying forms of life is called taxonomy, and it does two things: it names every animal, and then arranges them in groups according to their relationship. If all known forms were classified in a universally accepted system, and were named in accordance with the categories to which they belonged, scientists would be able to deduce certain characteristics about an animal (or plant) merely from its name, and would also be able to allot categories and names to newly discovered types. Such an ordered classification would be of immense value to zoologists and naturalists in particular, who are faced with the gigantic task of naming a bewildering array of living things that abound in the world. The work of taxonomists is never ended as life in all its variety cannot be expected to fit neatly into man-made patterns of organisation. Even the most fundamental division into the animal kingdom and the plant kingdom can be ambiguous; for there are some single-celled organisms which resemble plants in some ways and animals in others.

The only natural category in the animal kingdom is the Species: a

The scientific classification of three natives.

Classification	North Island Brown kiwi	Inanga (whitebait)	Stewart Island Short-tailed bat
Kingdom	Animalia	Animalia	Animalia
Sub-kingdom	Metazoa	Metazoa	Metazoa
Phylum	Chordata	Chordata	Chordata
Sub-Phylum	Vertebrata	Vertebrata	Vertebrata
Super-class		Anathostomata	
Class	Aves	Actinopterygii	Mammalia
Sub-class	Neornithes		Theria
Super-order		Teleostei	
Order	Apterygiformes	Isospondyli	Chiroptera
Sub-order	(Ratiti)	Salmonoidei	Microchiroptera
Family	Apterygidae	Galaxiidae	Mystacinidae
Genus	Apteryx	Galaxias	Mystacina
Species	A. australis	G. maculatus	M. tuberculata
Sub-species	A. australis mantelli		M. tuberculata robusta

unit which is reproductively isolated and generally breeds true, thus producing fertile offspring. Exceptions do occur, however, and such variants are referred to as mutants. Individuals within a species may differ in minor characteristics and occupy a distinct geographical area: these are defined as Subspecies. For example, the bush wren *Xenicus longipes* is a subspecies by virtue of distinctly separate populations in the North Island, South Island and Stewart Island, and named *X. l. stokesi, X. l. longipes* and *X. l. variabilis* respectively.

Species which are closely related form a single Genus. Thus the bush wren and the rock wren are taxonomically similar enough to constitute the genus *Xenicus*. The classification continues by grouping related genera into a family. *Xenicus, Acanthisitta* and *Traversia* (whose sole representative prior to extinction was the Stephens Island bush wren) collectively form the Family Acanthisittidae – the New Zealand wrens. And so it goes on; related families into an order, related orders into a class, related classes into a phylum and related phyla into the animal kingdom. Every living thing can be placed in such a classification scheme.

Scientific Names – the Binomial System of Nomenclature

In the middle of the eighteenth century a serious attempt was made to give permanent scientific names to all the known flora and fauna. A Swedish botanist, Carl von Linné, commenced the task and developed the system which still forms the basis of plant and animal classification. Latin was chosen because it was then the international language of science, and von Linné even Latinised his own name to Carolus Linnaeus. Linnaeus proposed that all organisms be classified according to general structure and shape, physiology, distribution and mode of life. This system gained widespread acceptance after the publication of his book *Systema Naturae,* in 1758.

The foundation of Linnaeus' naming of organisms is the Binomial System of Nomenclature which was invented a century before but not generally used. Every species is given a binomial (two names) which assign it to an existing genus (if it is sufficiently close to other members of that genus in most of its characters) and additionally gives it a unique specific name within that genus. The name is always printed in italics and gives first the genus, with an initial capital letter, then the species.

Environmental Adaptation

As well as being a catalogue of the names of living things, the modern classification is also a summary of their relationships. One hundred years after Linnaeus, Charles Darwin modified the system as a result of his theory of natural selection which explains how all animals and plants had evolved from previous forms of life. On the publication in 1859 of *On the Origin of Species* Darwin was able to devise a classification system which would not simply group animals according to their similarities or differences but would arrange them in such a way that showed their paths of evolution.

Darwin also noticed that many animals produced more offspring than ever survived, and that there must be some factor, such as shortage of food, which kept the populations at a stable level. There was, therefore, a 'struggle for existence' in the natural environment. He also observed that any variation which better suited the individual to its particular habitat would have two effects: to give the individual a better chance of survival, and, if offspring could inherit this characteristic, they too, would have a better chance of survival.

The other major points included in Darwin's theory were: firstly, that individuals within a population differ from each other in small but significant details (the concept of continuous variation) so that those animals best adapted to survive have an advantage; secondly, an individual which has the advantageous characteristics is more able to live and reproduce, so, given that hereditary resemblance between parent and offspring is a fact, those characteristics will be passed on; and thirdly, subsequent generations will gradually maintain and improve on the degree of adaptation realised by their parents.

The Origin of the Species was an immediate publishing success and Darwin's theory has stood the test of time. Others have strengthened the theory since, notably Gregor Mendel who founded the science of genetics, and Hugo de Vries, a Dutch biologist who extended the theory. De Vries showed that chance occurrences (induced by radiation or other means) can produce marked changes in the genetic make-up of an individual, and that these changes are inheritable. So if an individual is born with exceptionally sensitive hearing because of a genetic mutation, his offspring will probably have exceptional hearing too, because this characteristic is carried by the sex cells from one generation to the next.

Only small changes within a species could be accounted for with Darwin's theory, but mutations were needed to explain how the large changes leading to the establishment of new groups of animals could occur. So, in the 1930s, the modern theory of evolution (the Synthetic Theory) was proposed, which is based upon the theory of natural selection but includes the concepts of mutation and genetic inheritance.

The Habitats

Evolution is a continuing process, and even within the past 100 years this has been particularly evident in New Zealand. The wholesale removal of great tracts of native forest has changed the natural habitat of many native birds. Some, like the kingfisher and the falcon have adapted well and even benefited in the process; others like the robin and the wren have found it difficult to adapt and their numbers have dwindled alarmingly. In the forests of Rotorua the sambar deer has thrived and adapted to an arboreous environment after evolving for generations amongst the mountains of its Himalayan origins.

The study of living things in relation to their environment is called 'ecology'. As long as there is available space somewhere in the countryside, an animal or plant (or both) will occupy it sooner or later, and adapt to the prevailing conditions – temperature, rainfall, humidity as well as vegetation types and other animals, that are present.

The most basic division of the Animal Kingdom comprises those animals with backbones and those without.

This organised interrelation of life is called an 'ecosystem', where all the separate components function as a dynamically balanced whole.

A product of the enormous complexity of even simple-looking ecosystems is the extent to which animals and plants have come to rely on each other. The simplest type of dependence is the predator for its prey, or the parasite for its host.

New Zealand offers a wide variety of habitats to a relatively small variety of animals. Most countries of a similar size lack the variety and extremes of natural habitats which have evolved in New Zealand, yet host a greater wealth of wildlife. Small mammals, for instance, are conspicuously absent from a European viewpoint; butterflies are sparsely represented and freshwater fish, though locally present are often mismanaged and undernourished. However, most of the larger mammals are flourishing and the birdlife in many cases is unique. The native bats are our only terrestrial native mammals, our frogs are primitive and the tuatara is the sole survivor of a group of reptiles that became extinct 100 million years ago.

The conditions which govern the properties and peculiarities of a habitat are often arbitrary. The more specific the habitat the greater the number. With painstaking effort New Zealand could be subdivided into scores, perhaps hundreds, of habitat types each differing from the next in small but significant ways. The factors in determining a habitat type can be many and complicated but for the sake of convenience and simplicity six natural habitats have been chosen to cover the New Zealand scene. They are: the Seashore, Open Country, Bush and Forests, High Country, Inland Waters and Offshore Islands. Many creatures, especially insects, can be ascribed to more that one arbitrary habitat; conversely there are birds and deer species found exclusively within one habitat but totally absent from almost identical conditions elsewhere. Natural barriers like Cook Strait or the Southern Alps have long played a major role in the distribution of New Zealand's fauna.

THE ANIMAL KINGDOM

Invertebrates

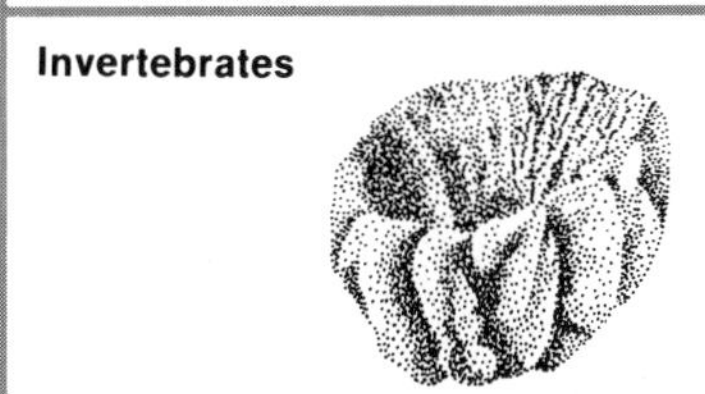

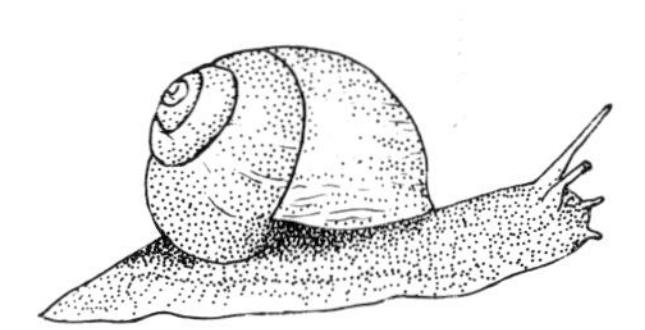

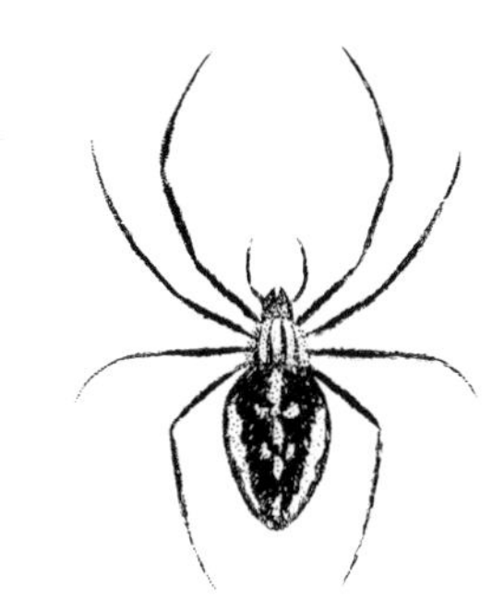

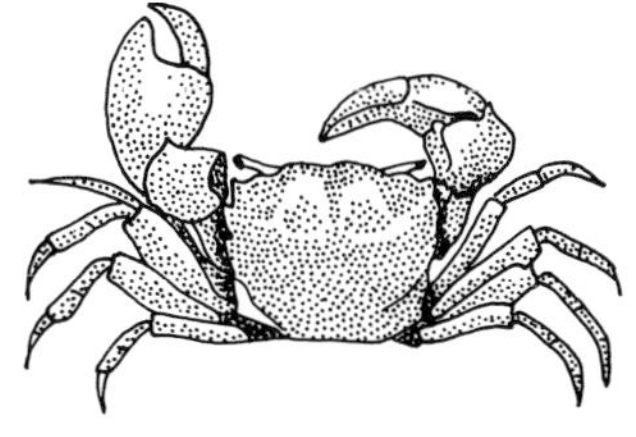

Animals without backbones fill 25 of the 26 phyla into which the animal kingdom is divided. The selection above of the more familiar invertebrates is representative of only four of the 25. Some phyla contain many classes: for example – **Arthropoda** consists of *Crustacea, Arachnida, Insecta* and ten other distinct classes.

Fish

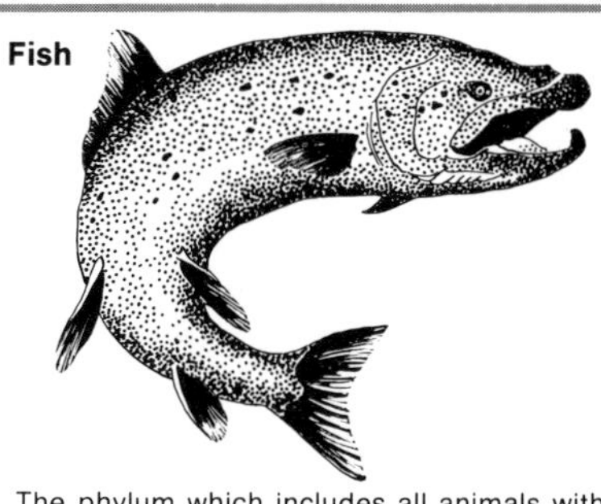

The phylum which includes all animals with backbones is called **Chordata,** and is divided into the groups represented in this column. There are three classes of fish, each as distinct from each other as birds are from amphibians.

Amphibians

Amphibians are the most primitive class of land-living vertebrates. The three orders in this class – frogs and toads; newts and salamanders; and caecilians, usually begin their lives in water to which most adults return to breed.

Reptiles

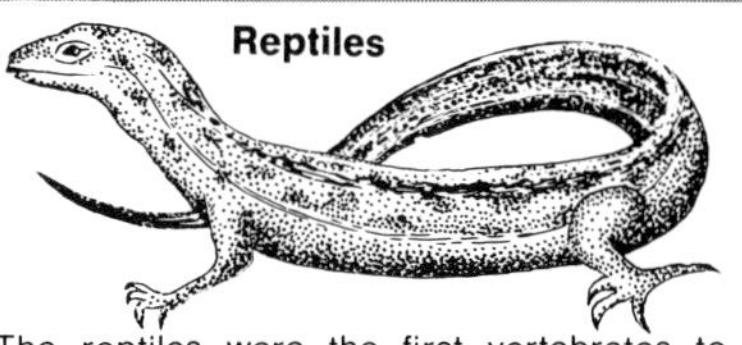

The reptiles were the first vertebrates to adapt to a wholly terrestrial life, and can be easily distinguished from amphibians by their scaly skins which prevent them from drying out. The four orders of reptiles are: crocodiles; lizards and snakes; turtles and tortoises; and the tuatara.

Birds

There are 27 orders of birds grouped in the class *Aves.* They differ from reptiles, from which they have evolved, by having feathers and wings and being warm-blooded. The largest order Passeriformes, consists of more than half of all known bird species.

Mammals

Typical mammals are warm-blooded, air-breathing vertebrates. There are three types – monotremes, which lay eggs; marsupials, whose off-spring are poorly developed at birth and are usually raised in the mother's pouch; and placentals, whose young are more advanced at birth.

SEASHORE

The ocean's edge, between the level of high tide and the level of low tide is the home of an abundance of strange animal forms exposed to greater extremes of environment than any other creatures on earth. Living a double life – sometimes submerged, sometimes in the open air – requires an exceptional adaptability, also unmatched in any other environment.

The periods of water cover enjoyed by the animals of the seashore vary according to their distance from the mean sea level. Tidal movements are related to the gravitational forces of the sun and moon. The water in earth's oceans is caused to 'bulge', and the 'bulge' moves around the globe in just over twelve hours. In New Zealand there is a lag of under one hour between the same tides on successive days. Life at the high-water mark may be underwater for only a few minutes every 12 hours, the period lengthening toward low water mark, where organisms will be submerged for most of the time.

Creatures of the intertidal zone, subject to the periodic covering and uncovering by the sea, experience severe physiological stresses. Exposed to the air in warm weather the organisms may even dry up completely; in winter the threat may be frost. Whenever left out of water, they will, in most cases, be cut off from food supplies. Shallow water will heat up or cool very quickly. Evaporation in warm conditions will result in a sharp rise in salinity and exposure to a deluge of rainwater can be dangerous to some organisms.

The creatures that live in this environment must, therefore, be able to deal with and survive these conditions as they occur. Just as important is the assault of the waves.

Most intertidal animals have thick calcareous shells that can be securely closed to create a relatively safe 'capsule' where conditions conducive to survival can be maintained. A barnacle, for example, is enclosed and protected from drying and wave action by a moisture-retaining shell made of six hard plates with a hinged cover of four more plates. When the tide rises, these open to allow six pairs of feathery legs to extend and sweep the water for the minute plankton on which the animal feeds.

Below: *The movement of tides – caused by the gravitational pull of the sun and the moon on the oceans – sets the limits to the zones of seashore life.*

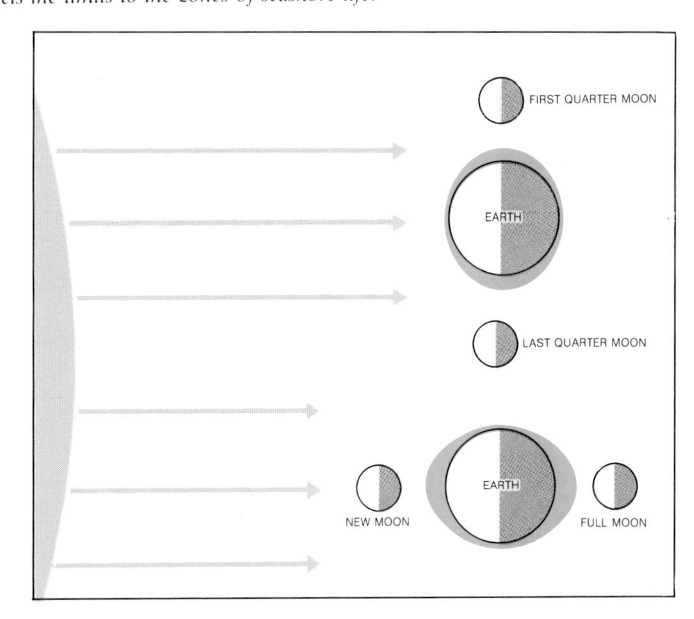

Shores composed of rock, sand or mud each have a distinctive marine community. The most fertile grounds are the rocky shores where pools, seaweeds and fissures provide shelter, food and shade.

Limpets are similarly enclosed in a resistant shell but are more highly adapted than the barnacle. The shell is conical and instead of being firmly cemented to the rock it is able to move by means of its extremely muscular foot. The gap between the shell and the rock allows it to feed when submerged by drawing water through the gap and filtering it in a similar way to the barnacle.

Animals not protected by a calcareous shell have developed hard or mucus-covered skins which reduce water loss by evaporation and also give some degree of thermal protection.

Basically, there are three types of seashore: rocky, sandy and muddy.

Rocky Shores

Rocky shores, usually exposed to the battering of wind and waves are probably the least hospitable, yet it is among the rocks that the permanent pools of the shoreline are to be found. Where shelter exists there is a great demand for the protection a pool or ledge affords and competition is severe. A crevice in the rocks may conceal a multitude of intertidal animals: crabs, sea urchins, barnacles, sponges, sea anemones, shrimps, oysters and starfish. Under stones and weeds and in pools, are found animals that are least able to resist exposure to air, such as fish like gobies, blennies and clinids. Clingfish attach themselves to rocks by their ventral sucker, and seahorses and pipefish live entwined amongst the seaweed.

A close look at the inhabitants of any low tidal pool that exists for only a few hours during each tidal cycle would reveal large assemblages of animals and plants. Algal growth is encouraged by the bright sunlight and lack of movement, especially in the shallower pools, and dense car-

The body and legs (cirri) of the adult barnacle are enclosed within a mantle made up of calcareous plates. The cirri are extended through an aperture between two pairs of these plates in order to catch food, and withdrawn as the tide falls.

pets of small algae are one of their most characteristic features. Many herbivores thrive on this rich growth, from small topshells to small sea-urchins and the large red shore crab.

Commonest of these algal grazers is the cat's eye snail, a fast moving species with a voracious appetite for the more delicate seaweeds. The scattered green-brown shells of this snail cover the bottom of most low tidal pools. Any sudden movement or passing shadow sends many of the shells tumbling and scurrying in confusion: these were empty shells that have been taken over by hermit crabs, alert and curious scavengers and predators prowling by the hundreds around tidal pools. The soft tail or abdomen of the hermit is protected by the borrowed shell and defended by the armoured legs and large strong pinchers that project from the shell aperture.

Also herbivorous are some sea-slugs, the small *Stiliger felinus* for instance, being a tenant of rock pools high on exposed shores. This creature is jet-black with a spindle-shaped projection forming its back.

Rock pools and intertidal ledges also house sea anemones which look like harmless plants, swaying gently. They are, on the contrary, carnivorous, their graceful stinging tentacles paralysing small fish.

Other specifically rock residents include the mussel, paua and oyster.

Mussels attach themselves to the rocks by spinning fine, strong threads which are secreted as a fluid by glands in the foot and are hardened by sea water. The threads function as 'guy ropes', holding down the mussel. Many bivalve molluscs cement themselves to the rock surface, like the barnacle. Limpets and chitons clamp themselves firmly with muscular feet, while other animals, instead of developing a secure method of attachment, find shelter on larger animals or in tiny cracks on the rock surfaces.

Three species of paua are found only in New Zealand. The largest, known as the common paua, is a velvety black animal which reaches a length of about 17 cm. Half its size, and bearing an attractive pink shell with a dimpled surface, is the southern paua. The virgin paua, smaller again, is found throughout New Zealand in many regional forms. All of these rock-dwelling shellfish feed on seaweeds or tiny plants that are rasped from the rock surface. In turn, they are regarded as a delicacy and the shells, especially of the common paua, have been commercially developed in the manufacture of distinctive decorative art.

Sea anemones are found from tidal pools to the abyssal depths, and are voracious eaters of any organic matter, living or dead.

Perhaps the most celebrated mollusc of more sheltered rocky shorelines is the oyster. In New Zealand, rock oysters characteristically occupy a zone about one metre wide just above the low water line. Like the limpet they are extremely tenacious, and protection is afforded by their two heavy calcareous valves tightly held together by powerful muscle fibres.

Of the 80 or so species of crab known to inhabit New Zealand waters, about 30 are likely to be seen in the intertidal zone, or washed ashore from shallow water. Many of these have very restricted ranges, Cook Strait possessing the largest variety. The common rock crab is, as its name implies, the species most often seen, living under rocks on semi-sheltered shores from the half tide mark up to high tide throughout New Zealand.

On more exposed rocky coasts dwells the large shore crab which can measure up to 7 cm across the carapace. When feeding it ranges widely but returns to crevices for rest or shelter. The red rock crab lives below the low tide mark but emerges at dusk to forage over exposed rocks. It is a brilliant reddish coloured species with spiny legs and body, and can grip the surf-lashed rocks with tremendous strength. Another common crab of the intertidal zone is the big-handed crab. Found under stones from the mid-tide down, and less than 25 mm across it is still conspicuous because of its dull yellow body, bright yellow nippers and red eggs.

Amongst the tufts of small seaweed in the rock pool, lives a unique insect, the marine caddis-fly. The larva builds an open-ended, horn-shaped tube, up to 10 mm long, to which sand grains and pieces of weed are attached, forming a complete camouflage. The adult fly can be found crawling on the weed patches.

In the highly saline rock pools washed only by spring tides, live the larvae ('wrigglers') of the salt-pool mosquito. During summer the large black adults darkly cover the surface of these pools.

The intertidal rock pool contains an amazing variety of plant and animal life.

The red rock crab (above) *spends the day below the low tide mark, emerging at night to forage among exposed tidal rocks. The brightly coloured firebrick starfish* (below) *clamps fast to a sponge on steep, rocky cliffs that plunge into deep water.*

Sandy Beach

In contrast to the rocky shore the sandy beach appears barren and devoid of animal life. In fact a great deal of animal life is present at low tide, but mostly below the surface of the sand, for such a smooth and featureless habitat provides little shelter. The only way in which animals can survive when the water recedes is to burrow. Sand is easy to burrow in, but it is difficult to prevent walls from collapsing and cutting off direct contact with the surface. To avoid the danger of suffocation, sand-burrowing animals have developed various ways of maintaining contact with the surface. Some worms form tubes that act as air shafts, and molluscs use their long siphons.

Even sand that appears dry still contains a great deal of water which exists as a layer around the particles, joining them together so that creatures below the surface do not dry up. At the same time, although contact with the air may alter the temperature of the uppermost layers, the effect penetrates no further than about 25 cm, resulting in a well-insulated environment below that level.

The New Zealand intertidal zone is home to many edible shellfish including the tuatua, toheroa, pipi, cockle and scallop. The cockle and pipi are just as likely to be found in mud as in sand, but both the toheroa and the tuatua prefer fine sandy beaches exposed to the full force of ocean waves.

Tuatuas are found on most beaches, usually on the lower part of the shore and it is a common sight around the New Zealand coastline to see scores of people wading in ankle deep water feeling for the mollusc with their feet.

Toheroas are not so abundant, occurring mainly on the west coast beaches north of Auckland with smaller populations near Wellington and the south coast of the South Island.

In isolated regions scallops are still quite common, preferring the shallow water below low water mark but often being washed higher up the beach. The scallop normally rests in a slight depression with the flat wave of the shell uppermost, and a thin layer of sand covering.

The ghost shrimp and the mantis shrimp, both inhabitants of the lower sandy shore, are adept at burrowing deep into the sand. The former is a filter feeder, agitating the water which it causes to circulate through its burrow and living on the tiny particles combed with its mouthparts. The mantis shrimp, which grows to over 10 cm long is a carnivore and frequently makes forays into the open to seek its prey.

To burrow, the fleshy foot of the toheroa is thrust into the sand surface; water jetted out through pores along the lower edge helps to puddle the sand and make penetration easier. Once buried the foot is dilated, wedging it into the sand; the shell valves are pulled rapidly together, forcing out more water, and the foot muscles are contracted, pulling the animal upright and down into the sand.

The intertidal rock pool and estuarine shrimp Palaemon affinis *is the most common and best known shrimp in New Zealand.*

Along the high-tide mark live myriads of small, insect-like crustacea collectively known as sandhoppers. These active creatures dig temporary shelters surrounded by small mounds of excavated sand. Since they require only a small amount of moisture to respire and prevent dehydration, they can survive easily at this level of the shore. After dark they emerge from their burrows to scavenge in the organic debris that accumulates around the upper tide mark and that provides an abundant supply of food. Masses of wrack – all kinds of seaweed washed up by storms – provide food for a host of scavenging animals which hasten its decay. Kick aside any of these patches on a sandy beach and amongst the scutter of disturbed creatures are sandhoppers, behaving like acrobats; isopods (slater-like) scurrying out of sight; rotund, pale brown darkling beetles, adapted for rapid digging, and dark, slender rove beetles, which are there to feed on the others. Kelp flies which lay their eggs on the seaweed sometimes rise up in clouds. Driftwood provides a stable enough shelter for the seashore earwig to brood her clutch of eggs and for the large (up to 7mm) curled grubs of the sand scarab beetle which feed on rotten wood and also on the roots of nearby marram grass.

Sandflies, though regarded as pests of both humans and domestic animals because of their blood-sucking abilities, are really not a serious threat as no human disease carrying species are known in New Zealand. There are 13 species in New Zealand, but only two species attack readily. It is the females that suck blood – for maturation of the eggs. The adults are most active during warm days, especially with high humidity, and can travel considerable distances with the prevailing wind.

Muddy Shores

At some point along the shore, mud and sand mix in varying proportions. Here the sand-dwellers and the mud-dwellers are inclined to merge, and occasional stones and boulders may also provide shelter and protection for animals more commonly found amongst the rocks. But where mud exists by itself, living conditions become more difficult than in other intertidal habitats.

Muddy shores consist of minute particles of inorganic matter, and often a substantial quantity of organic debris. The fine particles are easily carried in suspension in water and tend to block an animal's feeding and breathing mechanisms. Only animals which have adapted to this inhospitable realm are successful. These include bivalve molluscs like the cockle, the pipi, and the minute nut shell, univalves like the horn shell and mud snail, a few slugs and worms, notably the lug-

Mangroves are important primary producers in the biological food chain, harbouring a wealth of sedimentary food for certain animals. The abundance of deposit – feeding cockles, for instance, will attract carnivorous creatures like these whelks which prey upon the cockles.

Katipo

Mention the word spider in New Zealand and most people will think first of the katipo. Few spiders are harmful to man, but it is the few venomous species which are responsible for the general fear and mistrust. The katipo is not the only New Zealand spider capable of causing discomfort or pain, but seems to be the only one which is potentially dangerous. Although its favourite habitat is the foreshore, where it is locally abundant and most often seen under driftwood, fortunately it is a shy retiring creature.

The female katipo has a body length of 10 mm and is characterised chiefly by its shiny black globular abdomen – about the size of a garden pea – with a distinct red stripe down the middle. The male is much smaller. A second species, the black katipo, lacks the red stripe, and is found exclusively on the beaches of the northern half of the North Island. While only the fully grown female of these spiders is dangerous, it is advisable to leave well alone.

worm and the familiar, scuttling mud crab.

The mud crab is a most energetic burrower and the burrow is the centre of each crab's life, much time being spent maintaining and defending it. These crabs are extremely active at low tide, scampering about on the surface in search of dead animal material if available, or eating mud and extracting organic nutrients from it. Their acute eyesight allows them to detect movement over 50 m away and if danger threatens they dart to the safety of their burrows.

The lugworm lives at the bottom of a V-shaped tube with two openings in the mud surface. Mud or sand is drawn down the head shaft for food, and a continuous current of water moves through the tail shaft for breathing. At intervals the lugworm backs up the tail shaft to deposit casts on the surface.

One of the best known of New Zealand's shellfish is the tuangi, or New Zealand cockle. Though not a true cockle it is very similar in appearance but lacks the heavy banded ribs of the common purple cockle, and may grow up to 5 cm in width. It lives in large colonies just beneath the surface in muddy localities from mid-tide down.

Birds of the Shoreline

The Waders

Most of New Zealand's seabirds usually nest in colonies on outlying islands or steep and rocky coastal headlands; however, a large number, waders in particular, frequent the harbours, mud-flats, estuaries and sandy beaches of our shoreline. Of the 50 species of wader that may be found in New Zealand, about 40 exercise some form of annual migration. For some it is merely a passage between the North and South Island, for others a crossing of the Tasman to winter in Australia, but for most of the remainder it means epic flights to northern Asia or north-west America.

The wrybill is a typical domestic migrant. Those seen between the tides in the Auckland region in late December arrive in their hundreds from the South Island riverbeds east of the Alps. It is likely that many wrybills fly direct to Auckland, but a few do pause in the vicinity of Cook Strait or the east coast of Hawke's Bay. Come August or September the reverse journey is made to breed in the south before flying north to return faithfully to the same high-tide roosts.

Pied oystercatchers, though often breeding at fairly high altitudes in the South Island except Nelson and Westland, will then head for the coast. Many thousands spend the winter in the Manukau Harbour and the Firth of Thames, or further north. Three species of this wader occur in New Zealand: the South Island pied oystercatcher – a sub species found in many other parts of the world; the variable oystercatcher, which is found only in New Zealand, and the Chatham Island oystercatcher.

The variable oystercatcher, as its name suggests, has two varieties: it may occur in pied or black plumage, or in a range between the two. The pied variety can easily be confused with the South Island pied except that it is noticeably larger and the boundaries between its black and white areas are not as precise, one colour smudging into the next. White bars on the wings of both differ also, and recognition in flight is thus made a little easier. The variable's white wing bars are short and narrow instead of long and broad.

Not as restricted in its breeding, the variable oystercatcher will nest in coastal sand dunes the length of New Zealand and lacks the migratory instincts of its close relative. The totally black form is less common, though sizeable numbers are present in the Coromandel and Bay of Plenty where it outnumbers the pied form. The dominant form in the South Island is black rather than smudgy. Oystercatchers rarely feed on oysters, preferring worms and larvae that burrow in the sand or mud.

Most pied stilts spend the winter in northern New Zealand, but some, like these on the Otago peninsula, prefer to stay south.

The elegant pied stilt is quite common throughout New Zealand

The wrybill is known to breed only on the larger riverbeds of Canterbury and North Otago. From late December onwards the main population moves north in small flocks to spend the summer around Auckland shores. Notice the unique bill which turns to the right at its tip.

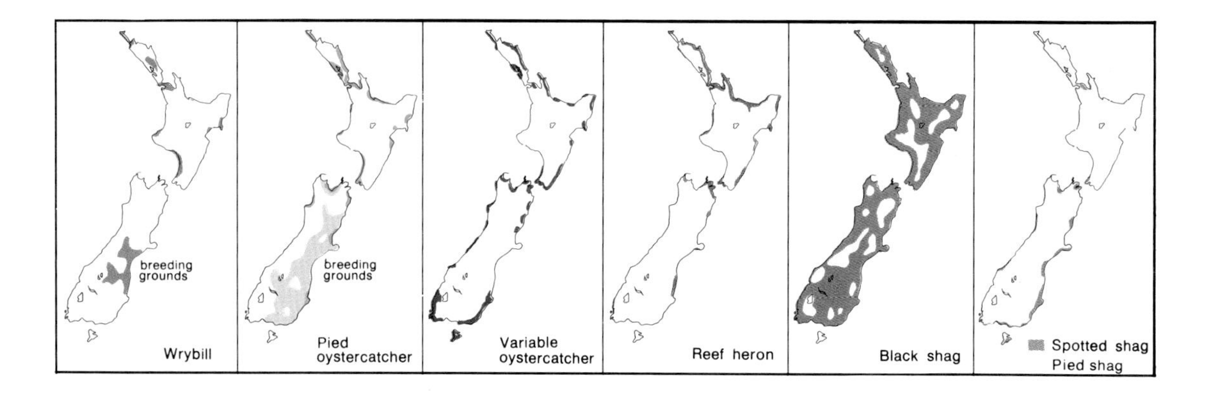

spending much of its time on swampy pastureland but as the paddocks dry out the birds move to the coast, usually travelling at night. Many South Island birds winter in the north, and there is some evidence of a northward dispersal of the rare black stilt whose numbers are probably fewer than 100, and whose breeding range is restricted.

Three species of dotterel breed in New Zealand, all members of the large family of plovers. They prefer bare or sparsely covered ground and sandy areas especially near water, although a distinct choice of habitat for breeding birds are the stony riverbeds of the South Island.

The banded dotterel is the most numerous and is widely distributed. After nesting, most birds gather in flocks at and near the coast, on our beaches, estuaries and wetlands in readiness to winter in Australia. Some stay behind to spend the winter in the South Island, but those that do leave, generally take off in a northerly direction and head west after recongregating in the vicinity of Parengarenga Harbour.

Distribution of the black-fronted dotterel is not so widespread and is centred mainly around estuaries and lagoons where the birds run smoothly and quickly along the water's edge. Both species associate in the river beds during the breeding season, nesting only a few metres from the water-line.

The third species, the New Zealand dotterel, is the largest. This approachable bird has a most peculiar distribution pattern. It is found in the northern part of the North Island where they choose the hot beaches and sandy hills, and on Stewart Island where they prefer the windswept tussock of the hilltops. Known to winter regularly in Awarua Bay near Invercargill, the southerners are only rarely seen between the Bay of Plenty and Foveaux Strait.

The very approachable New Zealand dotterel is the largest of the three species which breed in New Zealand.

During September the first of thousands of arctic waders arrive in New Zealand and quickly occupy their traditional stretches of foreshore from North Cape to Stewart Island. The migration continues throughout October and into November, the largest group by far being the eastern bar-trailed godwits from Siberia and Alaska. They are to be found, usually in flocks, on mudflats and coastal marshes throughout the country with concentrations in Auckland, Farewell Spit, Christchurch and Invercargill, the majority departing in late March.

After the godwits the most numerous of the wintering arrivals are the knots; then a mixture of turnstones, Pacific golden plovers, sandpipers, curlews and stints. All waders in New Zealand are strictly protected. In some surburban areas where estuarine land and tidal flats have been reclaimed, godwits now jostle with stilts and oystercatchers at their surviving resting places.

Gulls and Terns

Only three gulls are found in New Zealand, of which two are probably our most familiar coastal birds. The large southern black-backed gull is common everywhere, breeding in a wide range of habitats including lakes, mountains, and riverbeds as well as the coast. The smaller red-billed gull is also widely distributed though showing a marked preference for eastern shores and Cook Strait. Well known for its scavenging talents, this gull also catches live food from the surface of the sea often far from land. The other gull, less common on our shores, is the black-billed gull which breeds chiefly on South Island riverbeds and also at Rotorua.

A number of tern species may be sighted in the New Zealand region but only three are found on the mainland. Like the gulls, two species, the white-fronted tern and the Caspian tern – occur throughout New Zealand, and the third – the black-fronted tern breeds only on South Island river beds east of the alps. The largest member of the tern family, the Caspian tern, was once widely distributed but is now common only in New Zealand. It is bigger than the red-billed gull and possesses a heavier red bill, black feet and a conspicuous black cap. Most Caspian terns nest in colonies on secluded beaches or extensive sand dunes, and in winter large numbers assemble with waders in the northern harbours.

The highly gregarious white-fronted terns are a familiar sight almost everywhere along the coast. Sometimes known as sea-swallows due to their deeply forked, light grey tail, this species nests in a shallow scrape in sand or shingle, with varying amounts of grass and twigs, and often on the bare rock.

One of the lesser known terns, the fairy tern, is extremely rare; probably fewer than ten pairs breed annually among the sandy wastes of the Far North. Other members of the tern family are found on outlying islands like the Kermadecs and only rarely reach the mainland as stragglers.

Shags

Several species and subspecies of shag coexist in New Zealand, the best known being the large and sturdy black shag, which is distributed worldwide. Nesting in colonies near or directly over the water, this bird is typical of others in its family in that it has become so well adapted to an aquatic life that its wings can be used like giant fins to swim beneath the surface in search of food. Throughout most of the year the black shag is a solitary fisher despite its colonial breeding habits, but in late summer or autumn, large

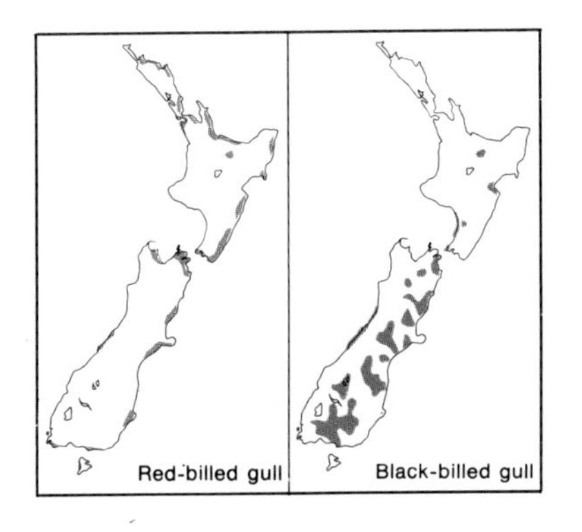

Red-billed gull guarding its nest. Breeding colonies of this species may number well in excess of 10,000 birds.

flocks gather in shallow tidal waters where there may be a seasonal abundance of shoaling fish.

Slightly smaller than the black shag, the pied shag has a white neck and underparts and a small patch of bare yellow skin on its face. It just as commonly nests on willows and pines above inland waters as in pohutukawas at the ocean's edge.

The little shag and the little black shag are about the same size, the former having a yellow bill instead of black, and white throat in the adult. Little shags also have a pied form. Although the most common of freshwater shags, both may frequent salt water lagoons. However, the little black shag is less common and has a range generally limited to the North Island.

Mainland New Zealand has one other common but unevenly distributed member of the shag family. Distinct in plumage and stature, the spotted shag is smaller than the black shag, with the upper parts greyish-brown, each feather having a black spot at its tip. The bill is longer and narrower than the other shags, and the feet are pale yellow. This species is a cliff-dweller, using every available ledge, crevice or sea cave to build a nest. A subspecies of the spotted shag is the blue shag, generally found on Stewart Island and parts of the West Coast of the South Island.
The rare king shag, once prized commercially and slaughtered for ladies' muffs, is found only on some small islets in Cook Strait.

Spotted shags nest in colonies on steep cliffs overlooking deep water. They are abundant in the South Island but only locally common in parts of the North Island.

Reef Heron

The only member of the heron family to specifically favour the intertidal zone is the dark slate-blue reef heron. Other herons are often seen near the water's edge but the reef heron may be distinguished by the thin white line on the throat, shorter legs and hunchback stance. Usually seen singly or in pairs from Northland to East Cape, especially in the vicinity of coral reefs, it is less common further south.The birds will feed on tidal flats close to their rocky home base, crabs and small rock pool inhabitants making up much of the diet. Their nests are made from a haphazard arrangement of sticks, sheltered where possible by an overhanging ledge, or in a cave.

Penguins

The real home of penguins is the sea and when they leave it to breed they come ashore on isolated coasts washed by food-rich currents from the Antarctic.

It is still uncertain whether the species that breed on the New Zealand mainland number three or four. Some ornithologists regard the white-flippered penguin as a local variety of the widely distributed blue penguin; others refer to it as a definite subspecies. Generally confined to the Canterbury coast and Banks Peninsula, the white-flippered penguin is slightly larger and a shade lighter than the blue, with white edges to its flippers.

New Zealand and Australia are the only places where the world's smallest penguin, the blue, is found. They are particularly vocal birds, especially at night, and they seem even noisier when nesting under the floor of the bach! Just about anywhere will do for a nest which is made of sticks and grass, placed under rocks, in burrows or caves or even in clumps of flax 250 m above sea level. They go off to sea in May or June and return to breed between July and November.

The yellow-eyed penguin is New Zealand's most distinctive penguin, the adult displaying a conspicuous yellow stripe above the eye which extends backwards to form a band right around the head. The young have no yellow band. Standing about 65 cm it is the largest of the mainland penguins. Like the white-flippered penguin, it inhabits the coast south of Banks Peninsula. Nests usually are a short distance from the sea in scrubland or bush.

New Zealand's other mainland penguin is the Fiordland crested penguin, found along the shores of Foveaux Strait and the South Island's west coast. It is a secretive bird shunning open spaces and spending much of the breeding season in the safety of coastal native bush.

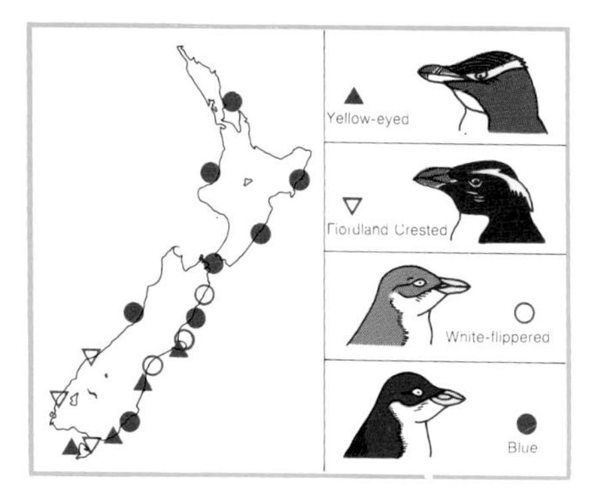

Above: *Of the eight penguin species that occur in New Zealand waters, only four actually breed on the mainland coasts. The blue penguin* (right) *of which several subspecies are separable, is the most vocal of all penguins.*

The New Zealand Fur Seal

This is the only species of seal that breeds on the New Zealand mainland, inhabiting rocky, boulder-strewn beaches of Westland, Fiordland and Foveaux Strait. Distinguished by its pointed nose and reddish-brown underfur for which it is the most prized of all seals, the fur seal was hunted almost to extinction in the New Zealand region. Happily, numbers are now increasing, and non-breeding colonies may be seen during the winter months at Kaikoura, Dunedin and Wellington.

New Zealand fur seals attain a length of 2 m – the females slightly smaller – and although some animals are present on the breeding grounds the year round, the males do not usually start defending their territories until mid November. Pups are born in late December, the cow remaining with her pup for a few days before spending a short time each day to feed at sea. On returning from the sea a number of calls are exchanged until the correct pup is eventually accepted.

Feeding of the pups by their mothers continues for nearly a year, almost until a new pup is born and one to two weeks after a pup's birth, mating takes place again. After all the females have mated there is a general dispersal away from the breeding grounds.

Despite its early slaughter, the fur seal is now increasing its colonies around the coasts of New Zealand.

Occasional seal visitors to New Zealand's mainland shores are the large leopard seal and the elephant seal. Both are southern residents, the leopard seal inhabiting the outer edge of the Antarctic pack ice and the elephant seal occupying most of the islands of the subantarctic region, but both have been seen on North Island and South Island beaches in spring or summer.

The common gecko is the most well known of the New Zealand geckos occurring under stones, logs and debris in cleared areas, as well as shingle banks within a few metres of high tide.

Lizards of the Shore

New Zealand is represented by 15 species of gecko and 21 species of skink. Both are to be found in a variety of habitats ranging from the seashore to mountain forest, but one species of gecko and three of skink are not uncommon living just above the high water mark. The common brown gecko favours large stones, seaweed and other debris where there is an abundant supply of food, throughout New Zealand.

Smith's skink is a coastal species found north of Gisborne on the east coast and north of Auckland on the west, and many offshore islands. Though sometimes found in pasture close to the shore its first choice habitat are the piles of driftwood and seaweed among stones and boulders immediately above the high water mark. Here it feeds on numerous forms of invertebrates common to such situations, basking on exposed rocks and large stones between meals.

A little larger but not as common, Suter's skink also inhabits the beaches of the northern part of the North Island, as well as many off shore islands. It is sometimes known as the black shore skink but this name is misleading as the dominant colour usually ranges from copper to dark grey with black markings on the flanks. Mainland populations are few yet it is the only native lizard which is totally restricted to the shoreline. Suter's skink is nocturnal, emerging at night to forage for insects that share the same habitat.

This moderately large species is New Zealand's only egg-laying skink. A clutch of four or five eggs is deposited in a nest scraped out of the sand or gravel or under a large stone. Hatching takes place in March and April.

The third skink that frequents coastal regions is the smallest of all our skinks, the decorated skink. This species is widely distributed throughout the North Island (Auckland, Bay of Plenty and Taranaki) and may be distinguished by its heavy black spots on chin and throat.

The shore skink is New Zealand's only egg-laying skink. It lives on stony beaches, often in seaweed or among boulder piles.

The Shores of the Hauraki Gulf

No natural boundaries describe the extent of the Hauraki Gulf. An imaginary line drawn from Cape Rodney near Leigh to the northernmost point of Great Barrier Island, and a similar line from Cape Barrier to Cape Colville would encompass about 7000 sq km, of which about a quarter are estuaries, shallow bays and relatively sheltered coastal waters. All kinds of natural (and modified) seashore habitats are represented in the near 1000 km of coastline that bound the maritime gateway to the country's largest city.

On sandy beaches along the mainland coast north of Auckland, exposure and surf action limit invertebrate life to burrowing forms, such as the tuatua and the swimming crab. Intertidal bands of hardy barnacles and the fascinating life of rock pools are to be found on the more exposed rocky coasts that separate the beaches and estuaries. In more sheltered parts wide reef terraces have been formed that provide paua, sea-urchins and the common octopus with holes and hideaways while softer mudstone or papa conceals a profusion of boring crustaceans, molluscs and worms.

A number of shallow harbours indent the Gulf coast with large expanses of mud-flat that support a great abundance – though less variety – of estuarine organisms. The upper reaches of the Firth of Thames consist of extensive mud-flats at low tide where pipis, burrowing worms, cockles and shrimps occur in vast populations.

Many kinds of fish live in the harbour channels, moving over the mudflats with the incoming tide, to feed. They include the flatfishes, mullet, parore and even snapper and small sharks.

The bird life along these shores is almost as varied as the marine life. The reef heron, white-faced heron, shags and terns are attracted by schools of fish in the shallow waters. Many species of resident and migrant waders feed and roost, especially in the Firth of Thames. Gulls continue to increase in and around Auckland as they take advantage of the untidy disposal of human waste. Even amidst the hurly-burly of the port the pied and little shags are no strangers.

Gannets, petrels, and shearwaters are frequent visitors and the occasional albatross may be sighted above the wave tops. On the many islands that dot the Gulf, certainly those north of Little Barrier Island, the archaic tuatara, geckos and skinks and the Polynesian rat, the kiore, still survive.

Since the first recorded nesting of white-faced herons in Otago in 1940, these graceful birds have rapidly become the most common breeding herons in New Zealand.

The white-fronted tern breeds only in New Zealand and is a familiar sight along most of our coastline.

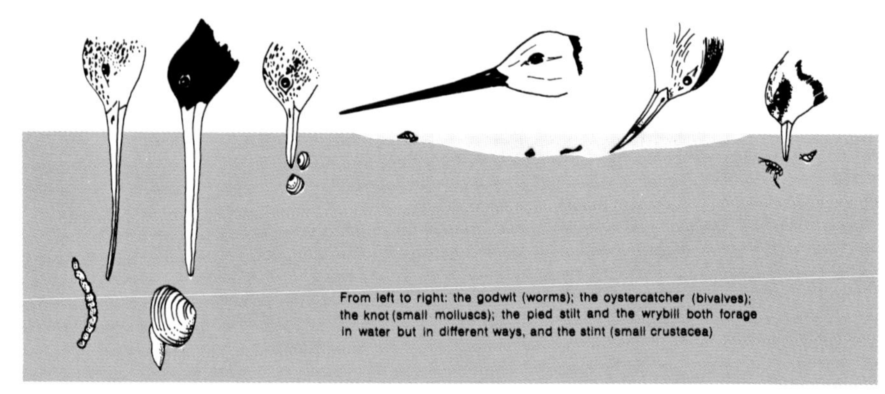

The food available to a wading bird depends on the nature of its bill. These closely-related species have evolved so that each can exploit different resources of a shared habitat. From left to right: *the godwit (worms), the oystercatcher (bivalves), the knot (small molluscs), the stilt and the wrybill (forage in shallow water but in different ways) and the stint (small crustacea).*

Tidal flats, like these at Miranda in the Firth of Thames, are rich feeding grounds for migratory waders.

Oblique coastal reefs on Auckland's North Shore form deep pools, even at low tide, which harbour a multitude of marine creatures like the sea slug Aphelodoris luctuosa.

OPEN COUNTRY

Once, certainly before the European colonisation of New Zealand, much of the North Island was covered by native forests, and huge tracts of the South Island were empty wastelands. Apart from National Parks and State-owned forests, most of the natural vegetation has long since been destroyed or intensely modified and replaced by either exotic forests or farmland. Land not devoted to agricultural pursuits is barren, alpine or urban.

Virtually all that may be called Open Country is farmland in some form or other, whether it be extensive grazing land for sheep or cattle, intensive cropping, orchards or market-gardening.

The wild animals that populate the open country are neither many nor varied. The wealth of small rodents and insectivores, for instance, that are found in other parts of the world do not occur naturally in New Zealand. Many of those that have been introduced, often for nostalgic reasons, are now regarded by farmers as pests. Unfortunately, this also applies to rabbits, mustelids and some birds and many insects.

The word 'introduced' is used for an animal which is not 'native' or 'endemic' and tends to suggest deliberate human intervention. However, many small animals such as rats, mice and particularly insects, have arrived uninvited, especially in the days before the importance of quarantine was realised. These are also referred to as 'introduced' although 'adventive' might be a better word. Such creatures are usually adaptable, invasive or aggressive and multiply prolifically.

With few exceptions the animals most familiar to New Zealanders are those of the open countryside. Hedgehogs and rabbits and rats have all adapted well and, despite highway casualities and organised destruction, are still abundant throughout lowland New Zealand.

The Hedgehog

Hedgehogs from Britain were released in the Christchurch area in 1869 and a little later near Dunedin. By 1915 they were numerous between the two cities and spreading fast. Some were liberated in Taranaki, Hawkes Bay and the Wairarapa between 1905 and 1910, and these too spread rapidly and quickly settled most areas of the North Island. By the 1930s hedgehogs had become so numerous that they were declared vermin, and acclimatisation societies offered a bounty of one shilling a snout.

Hedgehogs have adapted well to 20th century conditions and are abundant in city and countryside.

At birth, the hedgehog's skin is pink and covered with white pimples, but within an hour or two the skin dries and shrinks and the first spines begin to appear. Two more coats of darker spines emerge between the first coat and inside ten days the young are well protected.

The hibernation period for hedgehogs increases from north to south. In North Auckland they feed throughout winter, while around Wellington most hedgehogs hibernate between June and September. In the colder parts of the South Island their seasonal torpor may continue for six months. During hibernation, a hedgehog's body temperature falls from the usual 37°C to match the air temperature, sometimes down to 1°C. The heart may beat as slowly as once a minute, and breathing is at five minute intervals.

While feeding predominantly on slugs, snails, a variety of insects and spiders, and an assortment of grass grubs, caterpillars and cicada nymphs, hedgehogs have also been known to plunder the nests of pipits, pheasants and farmyard hens. They carry diseases such as mange and ringworm, especially in northern regions, but fortunately their own particular fleas were not introduced into New Zealand along with their host.

The Mustelids

Weasels, stoats and ferrets, liberated in various South Island localities in the latter part of the last century in an attempt to reduce the rabbit population, have since spread through much of New Zealand. Though each have quite different habitat preferences, their similarity may be confusing.

The weasel is the smallest carnivore in the world – about 25 cm long – but is not common in New Zealand. It has a short tail with no black tip, brown fur above and white below, and is so slim that it can run through a hole 2.5 cm in diameter. Weasels prefer the vicinity of old farm buildings, haystacks and grassland, wherever there are plenty of mice and small birds.

Stoats attain a length of 34-40 cm and the longer tail has a conspicuous tuft of black hair at the end. They are the most common and widely distributed of the three, and are found in both farmland and forests, in the mountains and city gardens.

The weasel is the rarest of the introduced mustelids though they have been seen occasionally in most districts except the southwest of the South Island.

The ferret is the largest of these mustelids, exceeding 50 cm in length. It is also stockier and its fur is thicker. Usually dark brown or black with creamy-yellow underfur, the comparative bulk and shagginess of the ferret helps to distinguish it from its smaller cousins. Ferrets live mainly on dry, open pastures and tussock country especially if sheep and rabbits are close by. They will eat whatever they can get hold of: birds, rabbits, mice, frogs, lizards and carrion.

Rats and Mice

New Zealand's four species of rodent – three rats and a mouse – were all introduced accidentally. Rodents are a very numerous and widespread order of mammals, characterised by having specialised jaw muscles and long incisor teeth which continue to grow and stay sharp by constant gnawing.

The kiore (Polynesian rat) and the ship rat are both discussed in other chapters, the former being mostly restricted to off-lying islands and the latter a dweller of bush and forest. The third species, the Norway rat, is the largest and has the widest distribution, being found in towns,

Rabbits spend much of the daytime sitting at the burrow entrance, seldom sleeping and always on the alert.

farms, rubbish dumps, swamps and in close proximity to rivers. Despite its name it is not a native of Norway; it originated in Russia and spread to Europe in the early 1700s.

Norway rats live in burrows and where they occur near rivers and swamps they are a menace to ground-nesting birds such as ducks, but are themselves preyed upon by the stoat. It appears that the population of Norway rats increases only in areas where the stoat is uncommon. Like other rats this species is a prolific breeder and the female may produce 50 or more young in a year.

The house mouse is widely distributed throughout the country, in towns, farms, forests and even the tussock and scree of high mountains. Since their arrival in the early 1800s they have spread even further afield and have now reached many of the outlying islands. However, their complete absence from islands where the species must surely have reached from time to time suggests that thriving Norway rat populations have thwarted the attempts of pioneer mice to establish themselves.

The Norway rat (above) *is the largest and most robust of the rodents in New Zealand, whilst the common house mouse* (below) *is the smallest. Both are prolific breeders.*

Rabbits

The European rabbit was first liberated in New Zealand in about 1839 but it took around 30 years to become well established. Then in the latter part of the century they increased and spread rapidly. In 1873, 33,000 rabbit skins were exported; in 1882, over 9 million.

The rabbit has been tolerated as a pest for more than 100 years because of conflicting attitudes – a valuable source of skins, or a pest causing a great deal of pasture damage. It has been estimated that ten rabbits require the equivalent food needs of one ewe. An attempt to reduce numbers through the disease myxamatosis was unsuccessful in New Zealand due to the absence of the rabbit flea which in Europe is the major carrying agent.

Unlike the hare, the rabbit is a gregarious animal, sometimes reaching densities of 50 to the acre. A colony consists of five to eight adults living in a warren, which can be a very complex network of branching burrows and escape tunnels at strategic points. Females are often included in the hierarchy; the subordinate males have little chance of mating and the low-ranking females produce few offspring.

Both bucks and does defend their colony's territory against other rabbits, the males marking the boundaries with dung heaps, urine and secretions of the chin glands rubbed on twigs or stones.

Feral Cats

The spread of rabbits in the late 1800s was accompanied by an increase in the numbers of wild cats, which even today appear to be more numerous where rabbits are plentiful. Probably the descendants of animals left by sealers and whalers in the early days, feral cats are now found throughout the country, both in open farming areas and forested land. Apart from rabbits, their diet consists of mice, rats and some of the larger insects, especially wetas and cicadas. Although they have a reputation for predating some birds, a natural balance seems to have been established.

Goats

The first goats to arrive in New Zealand were released on offshore and subantarctic islands by Captain Cook as a permanent source of food for castaways, and it is from these and other early introductions by

Goats are browsing animals and will eat any palatable leaves within reach as well as pasture.

missionaries, settlers and whalers that the present goat population has derived.

Feral goats in this country bear little resemblance to their noble Persian ancestors. Most populations have markings of black, brown and white, and various patterned combinations of these colours. In some parts of the country the populations recall the particular breed-type of the original liberation. For example, in the lower North Island ranges, goats resemble the improved British stock of the last century; in Taranaki they show distinct marks of the improved dairy breeds, and many of those in Hawkes Bay are obviously of Angora descent. Consequently there is a great variation in appearance, body size, weight and horn form.

Males and females both have horns but those of trophy size are restricted to the billy, sometimes with a spread in excess of 90 cm. There appears to be no set breeding season, and kids, quite commonly twins, can be seen with the nanny at any time of the year.

Goats show a preference for dry, sunny habitats but still manage to thrive in the wet conditions of the southern lakes region. When the weather is wet or windy – they vacate the open places in favour of the sheltered bush or scree beds.

Fallow Deer

Successfully introduced from 1864-1874, herds of this very attractive animal have now become well-established in at least 13 completely separate places. The largest and best-known are the Blue Mountain herd near Tapanui, the Lake Wakatipu herd, and the Wanganui herd.

Fallow deer are gregarious and seldom move far unless disturbed by hunting pressure. They also show a marked preference for lowland areas such as river flats and lower valley sides.

A mature fallow buck attains a shoulder height of 90 cm and a weight of about 70 kg. The doe is smaller and lighter, though there can be a wide variation of colour in animals of both sexes – from white to dark grey. The summer skin is light brown or fawn, sometimes with large white spots on the back and haunches. A dark line runs along the backbone to the characteristically long tail. The underparts are white or cream. In winter the colour changes to a much darker brown and the spots disappear.

Fallow deer have a well-developed sense of smell and extremely good eyesight but even so are extremely timid. When disturbed they run with a distinctive bounce due to all four legs lifting together, and are able to clear 2 m high obstacles with apparently little effort.

The antlers of fallow bucks are palmated, or flattened, towards the ends and the maximum size is reached at 7-9 years. Also characteristic is the proportionately larger tail than that of the red deer, the young of which are sometimes mistaken for fallow.

Lizards of Open Country

The common or brown gecko is

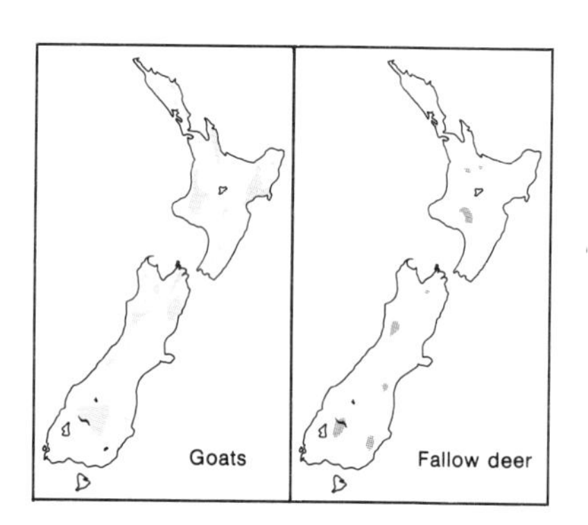

New Zealand's most abundant gecko. It is to be found in all types of open country from sandy coastal dunes to rocky outcrops in the hills. They are plentiful in Central Otago where large groups are often found together preparing to spend the winter in hibernation. Although nocturnal they may be seen in the open during the day basking in the sun.

Colour is variable in this species and patterns too can be transverse or longitudinal, often indistinct and in some cases appear to consist of scattered flecks and pale streaks. Generally however the background colour is a combination of brown, olive-grey or bluish-grey. It is more of a ground-dweller than other geckos which makes it particularly vulnerable to predating cats and rats.

The native skinks of New Zealand are all members of the genus *Leiolopisma*. Many can, with the aid of clawed toes, scramble about in small shrubs but they are best suited to life on the ground. Open country, usually uncultivated, with plenty of ground cover is typical of the habitat of many of our diurnal species. Some are quite common while others are rare, at least seven species being confined to offshore islands.

The common skink is to be found throughout New Zealand, from sea shore to suburban gardens and open country inland. It is generally brown with some variation in pattern and attains a length of 13-14 cm on reaching maturity. They are fond of basking but dive for cover at the slightest disturbance. All skinks in New Zealand are carnivorous, this species eating a wide range of invertebrates.

Above: *The common skink is found throughout the country in a variety of colour and pattern.* Below: *The large Otago skink is strongly diurnal and enjoys basking on the rocky outcrops where it lives.*

Widely distributed throughout the North Island is the decorated skink. It prefers open areas where ground cover consists of coarse grasses and rock-piles and is most active in the early morning or late afternoon. Light flecks embellish the coppery sides which merge gradually into a creamy-yellow under-surface. Their tails are lost easily, and a large proportion of those seen in the wild have a regenerated tail. Cats and other predators are no doubt the chief cause of this phenomenon.

Some offshore islands and a few mainland localities in the Bay of Plenty are the only known habitats of *L.moco*. This slender, medium-sized lizard varies greatly in colour and pattern, but is generally a dark coppery or olive-brown. A paler stripe begins at the snout and passes down the flanks to the tail, which is particularly long and finely tapered. It favours rough pasture where shelter is close and is fond of basking.

Three subspecies of the giant skink are peculiar to the South Island. The largest are heavy bodied, boldly patterned and reach an overall length of 30 cm. None are very common, each restricted to various small localities in the Canterbury-Otago region. The giant Otago skink, found only in Otago, is vividly attractive with back and sides a glossy metallic black, irregularly broken by yellow or pale lime blotches. It is strongly diurnal and enjoys basking on the rocky outcrops it inhabits.

Farmers' Friends and Foes

Many creatures that populate the soils and pastures of New Zealand's farming districts are regarded as pests, either because they cause disease in crops or because they compete directly with stock or man in order to feed themselves. Other creatures, notably ladybird beetles, bees and parasitic wasps are a positive asset to the farming fraternity.

A snail feeds by rasping the leaves with its numerous rows of teeth called radula. The common introduced garden snail bears up to 15,000 such teeth.

Slugs and snails are amongst the most obvious of horticultural pests. They forage at night and are a particular nuisance in wet weather. Closely related, the snail differs from the slug in that it carries with it its spirally-coiled shell into which the animal retreats when alarmed or at rest. The eyes of both are located at the tips of the larger pair of tentacles which can be extended or retracted at will.

Earthworms are generally considered to be beneficial as they assist in the breaking down of the soil, aerating it in the process. They are often found in huge numbers when the soil contains a high proportion of

Gorse weevils were liberated in 1931 in an attempt to combat the spread of gorse. It is now one of the most common insects in New Zealand.

organic matter. In New Zealand, earthworms vary in size from the tiny species that occur in the leaf litter, the larger species of the topsoil, and those of the subsoil that may grow up to 140 cm. This latter native species is restricted to the North Auckland peninsula and certain offshore islands. Its burrows are found on Great Barrier Island and continue down to a depth of 3 m and 20 mm in diameter. The worm surfaces very occasionally, hence its lack of pigmentation. However, those commonly encountered in pastureland and cultivated soils are introduced species which have displaced the greater variety of native worms which are now mainly confined to the bush.

Earthworms are hermaphrodite – containing both male and female organs – and when they pair one fertilizes the other, the eggs being deposited in a cocoon which is released into the soil.

Crickets

The black field crickets are widespread throughout the North Island and the northern part of the South Island. They are active at night and on warm summer evenings their strident chirping fills the air. These black, stocky insects are very fast movers jumping with the impetus of large hind-legs. They are fully winged but rarely fly. Several small native species do not develop wings.

Crickets are voracious, feeding on various kinds of herbage, and where conditions are suitable will seriously deplete pasture. They are shy creatures and shelter by day in cracks in the ground, under stones or in masonry. Plagues of the black field cricket have been known to invade houses and destroy furnishings, wallpaper and clothing.

New Zealand has one species of mole cricket easily distinguishable by their adaptation to a subterranean way of life. The hind legs are short and not adapted for jumping like their more common relatives, but the fore pair are remarkably modified like those of the European mole. The feet are splayed in a web-like fashion for constructing underground chambers where the insects live during the day and the females lay their eggs. Some species are winged but the New Zealand species is completely wingless. It is about 35 mm long and found in both undisturbed and cultivated ground. Northern Hemisphere species are sometimes injurious to cultivated plants but no such damage is recognised in New Zealand, and it is likely that they may even be beneficial as they feed mostly on grass-grub and porina larvae.

New Zealand's only species of mole cricket is completely flightless and lives a secret, subterranean life.

Katydids and Grasshoppers

The bright green katydid has a similar distribution to the black field cricket but is less of a pest. Slow moving and clumsy in flight, it lives in foliage where its leaf-like wings provide good camouflage.

The katydid is the only New Zealand member of the long-horned grasshoppers to prefer gardens where occasionally it will cause damage to fruits and flowers. The other three New Zealand species inhabit the longer grasses of tussock and sedge at higher altitudes.

The locust is the only fully winged short-horned grasshopper in New Zealand, and can often be seen in open, wild areas such as sand dunes. This infamous species is capable of forming vasts swarms which, in some parts of the world, do considerable damage. The migratory locust exists in two phases, solitary and gregarious. Fortunately the conditions which cause the latter phase do not occur in New Zealand, though the common lowland grasshopper will occasionally congregrate in sufficient numbers to be regarded as a pasture pest.

Beetles

Over 4,500 species of beetle are found in New Zealand in just about every kind of habitat imaginable. Like many of our other insects, the beetles have suffered greatly from modification or destruction of native vegetation. Only a few endemic species have adapted to cultivated land and pasture, so those most

The lace-like wings of the katydid provide excellent camouflage when sitting motionless in the foliage, but the adults are conspicuous and clumsy in flight.

The two-spotted ladybird is a predacious beetle and is well established wherever aphids and scale insects are plentiful.

likely to be found on farms or in gardens are the introduced ones. Specifically introduced to control pests such as scale and aphids are several species of ladybirds. As well as the familiar red and black varieties, we have the beautiful steel-blue ladybird commonly seen on citrus trees where it is a predator on scale insects.

Tiger beetles are insatiable predators and both larvae and adults inhabit almost any clay bank which is sparsely covered and exposed to the sun. The larvae construct narrow tunnels where they lie in wait for their prey. Supported by their legs and a pair of prominent dorsal hooks, they seize any unsuspecting insect that happens along before dragging it down the tunnel to be devoured. The adults are swift runners and will take off for short flights when disturbed.

One of the largest beetle families is Scarabaeidae which includes the cockchafers, scarabs and dung-beetles. They comprise some of the most significant agricultural pests in New Zealand. Grass-grubs, the larvae of chafer beetles, are fairly typical of the family, destroying untold acres of lawns and pasture by eating the roots. This is an example of native insects whose natural grassland habitat has been vastly extended by the work of man.

The most common of the grass grub adults is the brown beetle which lays its white globular eggs in the top 15 cm of soil. These hatch after 10 days or so and the young grubs immediately start feeding on

The brown or grass-grub beetles have become significant agricultural pests because they have been cultured by man's artificial development of pasture. Their natural parasites cannot survive under such unnatural conditions.

the plant roots, continuing throughout the winter months. In early spring they burrow deeper and form earthen cells in which to pupate. By October the adults begin to emerge, often thousands at the same time, to feed on leaves of trees, shrubs and vegetables which in an epidemic year can be completely defoliated.

Butterflies and Moths

Probably the best known endemic butterfly is the red admiral which is quite distinct from the red admiral species of the Northern Hemisphere. Common throughout the country it may be seen at any time of the year. Being an amazingly strong flyer, it has established a subspecies on the Chatham Islands and has even reached the coast of South America. The closely related yellow admiral is self-introduced (adventive) and seems to be on the increase. The caterpillars of both species feed on stinging nettles.

The common copper is as widespread as the red admiral but its colour pattern is extremely variable. Basically, the wings are a bright coppery orange, bordered and veined in black. Its larval food-plant is the straggling pohuehue vine.

A familiar sight to most New Zealanders is the handsome monarch butterfly. A native of North America this orangy-red, black-veined insect was once only a rare migrant to these shores, but as suitable foodplants (swanplant in the milkweed family) for the caterpillar became established, this beautiful species spread rapidly throughout suburban gardens and open country.

The life history of the monarch butterfly is typical of the metamorphosis that takes place in a great number of insects. First the female lays her eggs on the leaves of the foodplant; the eggs hatch and the green, white and black striped caterpillars eat almost constantly until, in a few weeks, they are fully grown. Each caterpillar will then hang by its rear end, moult its skin and change into a pupa or chrysalis, a glossy pale-green casket adorned with a few brilliant gold spots. Inside, over the next few days, the most astounding change takes place, and the casket, now transparent, splits open. The adult butterfly emerges, stretches its legs, pumps air into the veins of its wings and flies away.

Following its discovery at Napier in 1929, the pestilential cabbage white butterfly is pervasive wherever brassica crops are grown. However, it is partially controlled now by introduced parasitic wasps whose larvae devour the caterpillars and pupae.

New Zealand's most plentiful native butterfly is the small common

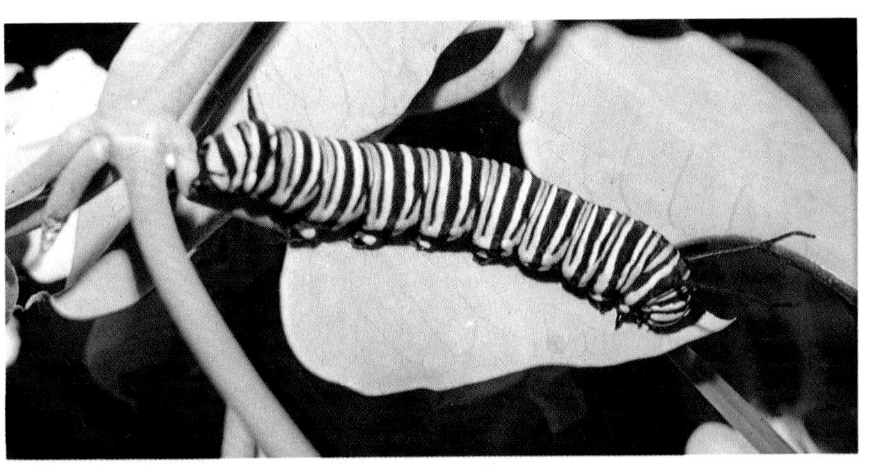

The caterpillar of the monarch butterfly will devour as much of its foodplant as possible prior to pupating.

Some of the more common butterflies of open country and suburban gardens.

blue, which appears on the drier grasslands and dunes. In late summer they may be seen flitting over stony riverbeds or along roadsides. The male is a bright sky blue on the upper surface of its wings; the female a drab brownish-grey.

A common visitor to New Zealand is the Australian painted lady. As its name suggests this is a brightly coloured butterfly with a 5-6 cm wing-span, but although it has been recorded in most districts it is seldom abundant and probably does not breed here.

The family of moths (Hepialidae) that includes the large puriri moth also contains a number of species responsible for serious pasture damage. These are the porina moths: primitive insects lacking functional mouthparts, and with a similar pattern of veins on all four wings.

Porina caterpillars, denude the pasture but, unlike grass-grubs, do not eat the roots. The moths are heavily built with furry bodies and are often seen on windows and around street-lights on summer evenings.

Perhaps best known of the introduced moths is the large and beautiful gum emperor. Since its accidental arrival – probably as eggs – in Wanganui, it has spread throughout the North Island and in the South Island wherever gum trees abound. The strikingly colourful caterpillars can grow up to 12 cm in length before they spin their brown cocoons of hardened silk on the leaves or twigs of the host gum. After wintering as a pupa, the adult tears its way out by means of a special hooked spine on the fore wing margin. The male is about 10 cm across the wings and has very feathery antennae; the female is larger. Both are pinkish-grey, with dark transverse lines and prominent black, orange and pink eye-spots on all four wings.

The convolvulus hawk moth, also known as the sphinx or kumara moth, is another introduced species, not as colourful but almost as conspicuous. The caterpillar may be either greenish or yellow-brown, and is easily recognised by oblique lateral stripes and a large reddish horn on the last abdominal segment.

Since its arrival from Australia the gum emperor moth has spread throughout the drier parts of the country, expecially where eucalypts are plentiful.

It was a pest well known to the pre-European Maoris who attempted to control it by burning kauri gum in their kumara patches. The wings of the adult moth are long and pointed, of approximately 9 cm span, streaked with dark brown. Alternate pink and black bands ring the heavy, furry abdomen.

Bees

Before the Europeans colonised New Zealand there were no honey bees, no bumble bees, but 40 kinds of native bees found nowhere else in the world.

Honey bees were introduced from 1839 and were soon established throughout the country. The hive of the honey bee is familiar to most people, consisting of one queen, several hundred drones and thousands of workers, all living in hexagonal bees-wax cells. The queen is the largest and her only function is to lay eggs – up to 3,000 a day. The drones, which have no stings, are the males and their sole occupation is to mate with any new queens to maintain the continuity of fertile eggs. Worker bees, as their name implies, do all the work. They gather in the pollen, build the honeycomb and feed all the larvae that hatch from the queen's eggs. And with their stings they defend the hive from other animals.

Since 1839 when honey bees were first liberated they found the new land to their liking and were soon established throughout the country.

Bumble bees were liberated near Christchurch in 1885 to improve the pollination of clover. It was noted in England that red clover produced good crops of seed only when there were plenty of bumble bees about. The reason: because red clover flowers are very deep and the nectar at their bases was out of reach of the honey bee's tongue. The longer tongue of the bigger bumble bee allowed the nectar to be collected and in so doing the flowers were frequently pollinated, thus providing better seed crops and greatly increased yields. Bumble bees do not produce honey but they have been of great value as pollinators.

The Case Moth

The male of this unusual insect is almost black and measures about 3 cm across the wing tips, but the adult female is a degenerate creature, being without legs or wings. The head and thorax are very small. In fact the insect is somewhat like a large maggot. She never leaves the protection of the silken bag and in mid-summer lays eggs within it. Immediately after hatching, the young larvae leave the case and roam over the host plant. In three days or so the young caterpillar constructs a conical, elongated case which it carries about with it in an almost upright position. Later the case becomes large and rather unwieldy and is dragged along in a horizontal position, or it may hang head downwards.

There are two apertures in the case: one large one in front from which the head projects, and a small one at the rear end through which waste falls. The young caterpillars may be found in late summer or autumn, throughout the country. Favourite foliage is manuka, broom, willow, radiata pine and macrocarpa.

Birds of Open Country

With few exceptions the land birds of fields and pastures, gardens and lowland scrub are introduced species comprising mainly song birds and game birds. The birds whose habitat is governed by the proximity of water are discussed in another chapter. These are the ducks, rails, waders, herons, grebes and others which restore the balance somewhat and are mostly natives. The reason for this is obvious: the greater part of the country, particularly the North Island, was once adorned by bush and forest, and most natives have simply not adjusted to an open country environment. Finches, larks, starlings, magpies, blackbirds, thrushes and pheasants are part of the country scene, and often the only birds in sight.

Blackbird and Song Thrush

The natural distribution of both of these birds is almost worldwide, but it required the early settlers in New Zealand to notice the lack of familiar songsters before they were imported.

Both species can live in close proximity because of their different diets and therefore do not compete seriously for food. The blackbird is probably the dominant species and the bolder pioneer, moving higher into the hills.

Sparrows, Finches and Buntings

The hedge sparrow and the house sparrow, though similar in size and name, belong to two different families. Both were early introductions and both are now common, and that is where the similarity ends. Though widespread, hedge sparrows are not particularly well known. Their habitats range from parks and gardens, salt marsh and mangroves, farmland and scrub and even high into the mountains. Their pleasant warbling is perhaps noticed more than they are themselves.

A male blackbird feeding its sole remaining fledgeling. Two to five eggs are normally laid between July and December.

House sparrows prefer built-up areas and are seen less frequently in rural settings. These drab and cheeky little birds have followed man to every corner of the globe and despite the dangers of city living and domestic cats, still flourish in great numbers.

The finches are well represented in New Zealand and have adapted well to local conditions. The ubiquitous chaffinch is the best example having penetrated the lowland bush, orchards and even montane beech forests. The male is a most attractive bird, the female rather drab, but both carry distinctive white bars on the wings which makes identification, even in flight, quite easy. The goldfinch, easily the most beautiful of the introduced songbirds, is remarkable in other ways too; an exquisitely sweet song, a beautiful nest made from thistledown, cobwebs, hair and feathers, and superb eggs of a delicate bluish-white with red blotches at one end. It is not surprising they were once very popular as cage-birds in Europe. Since their liberation in New Zealand they have so prospered that they are now more common here than in Britain, from whence they came.

The nest of the goldfinch is usually 2 or 3 m above the ground in fairly dense green foliage.

Greenfinches are quite plentiful throughout the country up to 600 m, and as common in suburban gardens as they are in open country and exotic forests. The conspicuously heavy bill indicates an ability to eat large, hard seeds and berries. Bright yellow markings on wings and tail add to the striking appearance. Greenfinches flock together in autumn and often head towards the coast where they may be seen feeding amongst other birds on the foreshore.

Another early introduction was the redpoll, the smallest of the finches to make a home in New Zealand. This bird has been amazingly successful and is as numerous above the bushline as it is among the grasslands and wetlands. The male breast has a variable amount of pink and both sexes sport a crimson blotch on the forehead; otherwise their appearance is rather sombre.

Buntings are closely related to finches and the most familiar of them is the yellowhammer. These birds, about the same size as the house sparrow, are a common sight in open country, often perched on roadside posts or fossicking for seeds in crops and grasses. Both sexes have prominent outer tail feathers of white; the male's head and breast is a bright yellow but the female is again somewhat subdued. A similar bird, but restricted mainly to the east coast of the South Island, is the cirl bunting. The black throat of the male distinguishes it from the yellowhammer though the females of each species are very difficult to tell apart.

The jaunty introduced and now well-established starling has a cheerful song, occasionally imitating the pukeko, fantail or California quail.

Starlings and Mynas

A wide contrast of attitudes exists toward these two birds. The starling is one of the farmer's few friends, consuming great quantities of noxious grubs and insects. It is one of the most abundant birds around the world and highly gregarious. Although extremely common throughout New Zealand it appears that rivalry between them and the mynas in the northern half of the North Island is allowing the latter a gradual numerical superiority. The sometimes raucous myna is certainly the most common roadside bird of the north and has the reputation of being an aggressive scavenger, and a raider of the nests of other birds.

However, although it will sometimes compete for sites with other hole-nesting species, the fact that a myna's diet consists largely of pasture insects, makes it a useful ally.

The Pipit and the Skylark

The New Zealand pipit is one of the few native birds to have benefitted by man's clearing of the bush, and with more open country available it has increased significantly in numbers. The skylark, superficially very similar to the pipit, came to this country in 1864; both are plentiful and widespread.

They are more easily distinguished by their behaviour than by physical characteristics. Only the skylark has a crest; the pipit is a duller shade of brown and its light eye stripe is more prominent. Both walk and run along the ground but only the pipit will spasmodically wag its tail. In the air the skylark has few peers as it hovers so high, often out of sight but seldom out of hearing.

Self-introduced from Australia the welcome swallow has adopted the open country of New Zealand in spectacular fashion and is now found from North Cape to Bluff.

The Welcome Swallow

Open country, abundant food and negligible competition have favoured the continuing spread of this attractive bird of the Pacific since its first recorded nestings here in 1958. From being occasional visitors in the Far North they are now well known throughout the country in suitable localities. Their preferred habitat is farmland where there is plenty of water – streams, ponds, even saltwater inlets.

The slender, streamlined body, forked tail and long pointed wings are all very noticeable when feeding on insects taken on the wing. Their flight is swift and graceful, sweeping, darting, banking and skimming close to the ground or water, then swerving skywards to circle and wheel overhead.

Nests are made usually in close association with man; under farmhouse eaves, in barns or sheds, under jetties or bridges. Mud pellets, reinforced with grass or roots, form the deep nest before it is lined with dry grass, feathers, and hair. Swallows often return to the same nest year after year.

Rooks and Magpies

The rooks of Hawkes Bay, southern Wairarapa and Canterbury are the descendants of European stock brought to New Zealand in the 1860s. They are large birds, glossy black with a bare face, and a slow, deliberate wing-beat. Large flocks roost together in tall trees in the neighbourhood of newly ploughed land or pastures. Farmers have never been quite sure whether they are harmful or beneficial.

Two forms of magpie have been introduced from Australia; the white-backed which is common all over New Zealand, and the black-backed, restricted to Hawkes Bay in the North Island and the Kaikoura – north Canterbury region in the South. Both magpies are slightly smaller than the rook but their voice is far more melodious. That of the black-back is a harsh bell-like song, but the voice of the white-back has been described as 'a glorious carol'.

The Little Owl

Introduced into Otago from Germany in 1906 in an attempt to control the small birds which were raiding orchards, the little owl has gradually spread over most of the South Island east of the Main Divide. It is the smallest owl in New Zealand and unlike the others it is not strictly nocturnal, being frequently heard during the day and seen perched on a post or branch, enjoying the sun. Its flight is low, rapid and undulating. Despite its size the little owl is a hunter, preying on mice, insects, small birds and lizards, usually caught at dawn and dusk. No attempt is made to construct a nest; a hole in a tree or building, or a disused rabbit burrow is sufficient comfort for the three or four nestlings hatched in November.

Australasian Harrier

Of the two hawks to be found in New Zealand the harrier is by far the more common. A bird of the open country, it is frequently seen gliding over hills and paddocks with wings forming a shallow V as it scours the land for prey or carrion.

The aerial antics and sharp cries in early spring signify the beginning of the harriers' courtship display. Males will try to impress the females by spiralling, swooping flights and graceful recoveries. As with many birds of prey, the female is larger than the male, but it is very difficult to tell them apart in the field. The nest is large, usually built on a platform of sticks and tussock well above water level on swampy ground. A shallow depression is made for the eggs.

A variety of food makes up the harrier's diet, mainly carrion, frogs, mice and insects, but like the native falcon it has been known to kill the odd hare – an animal more than five times its own body weight.

Commonly confused with the skylark, the pipit is slightly larger and only occasionally will it be heard on the wing. The skylark's aerial song is constant and unmistakable.

Pheasant and Quail

Both of these, introduced as game birds during the latter half of the last century, have taken well to the New Zealand countryside, especially in the North Island. They belong to a family which also includes grouse and partridge and have long been admired for their attractive plumage. The present population of pheasants is a mixture of several subspecies, notably black-necked, Chinese ring-necked and the Mongolian.

They have adapted successfully to farmland and scrub-waste, and in some northern parts to lupin-covered sandhills also. Their populations and distribution are much reduced as a consequence of intensified farming as patches of gorse and other low cover are cleared.

Like the pheasants the three species of quail liberated in New Zealand are predominantly terrestrial, feeding and nesting on the ground. The California quail and the Australian brown quail, were introduced subsequent to the decline of native quail – the koreke. About 1900, after the koreke became extinct, the Virginia quail, another native of North America, was brought in. Nowadays, California quail are distributed throughout open country in both main islands where escape cover is close at hand. The male is a handsome bird, mostly grey, possessing a black face edged with white and a plumed crest conspicuously larger than that of the female.

The brown quail is to be found in the North Island only, and being smaller and preferring denser vegetation, is not so well known.

Both the pheasant and the quail have similar feeding habits relying

The harrier hawk has benefitted considerably by the felling and burning of forest and bush and has become a common sight throughout the country.

The little owl is a bird of pasture-land in the South Island and is frequently seen during the day perched on a post or tree-stump enjoying the sun.

Pheasants, like this black-necked species, were among the first birds to be introduced. Most pheasants in New Zealand today are a result of cross-breeding with other subspecies.

mainly on herbs and grasses, grain, berries, seeds and insects. The larger pheasants occasionally include lizards in their diet. A characteristic of game birds is their need for a frequent intake of grit and small stones which enable the gizzard to grind up food, especially such hard morsels as maize and acorns. In country deficient in suitable natural grit, pheasants are attracted to the roadside in search of it.

The Spur-winged Plover

This bird, between tui and pigeon in size, is an Australian immigrant that has spread northwards from the south of the South Island. It has become firmly established south of Timaru and in many parts of Westland and Canterbury; the coastal population north of Wellington and the odd sighting near Gisborne suggest that it is continuing to extend its range.

The spur-winged plover is the only representative of the lapwing plovers found in New Zealand. It is easily recognised with its bright yellow facial mask and its sharp projecting wing-spurs. When in a threatening mood the bird displays its wings so that the spurs protrude aggressively. The spurs are also used in flight when defending its territory against marauding predators such as the harrier.

This is a bird of the open country, with a preference for wet places. It may frequent beaches and estuaries or stony, dry inland areas, but the typical habitat is damp and open with short cover. The bird, and its eggs, are fully protected. Because it consumes many insect and larvae pests it is generally useful to farmers.

The largest of the plovers to breed in New Zealand, the spur-winged plover builds a well-camouflaged nest on dry stony ground usually near a swamp.

BUSH AND FOREST

Millions of years of isolation; the nature of the vegetation that covered New Zealand when it was connected with other land masses; evolutionary trends of the plant world since; the changing geological structure and climate – these are the circumstances that have given New Zealand its distinctive forest cover and which have affected the animal communities of the forests.

Fossil evidence indicates that the characteristic New Zealand bush, with its dominance of podocarps (rimu, miro, totara) and southern beeches *(Nothofagus)*, may be traced back to the Cretaceous period of 136 – 65 million years ago. The country's ancient past and its lengthy geographical isolation is reflected in the similarity of its forests to the modern forests of southern South America, the likeness extending to the community structure of the forest fauna. This isolation together with a temperate, protective forest roof has meant that the forest life has had little need to adapt.

Prior to European settlement the land was already fractured with considerable erosion resulting from fire, high-intensity storms and earthquakes. The devastating fires of the early settlers, both Polynesian and European, altered much of the lowland areas and where repeated burning took place there was little regeneration.

It is difficult, therefore, to reconstruct the patterns of early forest distribution in some areas. In the northern region little of the original kauri or podocarp forests remains and while pockets of regeneration remain, most virgin forest has disappeared. On the west coast of the South Island podocarp beech and mixed-podocarp/beech forest must have originally formed an almost complete cover from the sea to the timber line. Except for limited strongholds such as State Forests, National Parks and other reserves, podocarp forests have been cleared for farming.

At the time of the arrival of the Maori the greater part of the North Island was forest-clad, dark, dense and almost entirely evergreen. Except for the occasional rain-shadow region the temperate climate gave ample rainfall that in turn created luxuriant and complex forest growth. Swamp forests were extensive along the lower reaches of most rivers, and conifers dominated exposed ridges. The forest floor was densely covered with ferns, mosses, liverworts and layer upon layer of decaying vegetation, presenting the appearance of a tropical rain forest.

The forest undergrowth remained complex and dense only because of the absence of browsing and grazing animals. New Zealand was unique in this respect, but the importance of this was only realised when introduced animals penetrated the forest. Two species of bats were – and still are – the only mammals native to these islands, thus the non-existence of predators resulted in the development of a rich bird life.

Today the remnants of indigenous forest still exist largely occupying the more inaccessible and steeper regions, while forests of exotic trees produce most of the country's timber requirements.

Of the forested land in the North Island 65 per cent is still indigenous and 35 per cent exotic. For conservational and recreational purposes some small native shrubs and tree ferns are retained in the larger exotic plantations.

There is a noticeable difference between the forests of the North Island and those of the South Island. Exotic plantations in the south are not as large as they are in the north and indigenous forest is less varied often being dominated by one or two species of beech. Rain forests composed of podocarps, kauri and broad-leaved hardwoods prevail in the north, and beech is mostly confined to the main mountain ranges. In the South Island all beech species are absent from the central West Coast and Stewart Island but nearly all other South Island forest is either a beech-podocarp mixture or pure beech.

The mixed beech-podocarp forests of the foothills and middle altitudes have become the major resource of indigenous timber. They consist of red beech, silver beech, hard beech or mountain beech in varying combinations, usually with scattered associations of rimu. At higher altitudes and on poorer soils the species become fewer, the less hardy ones diminish and resilient mountain beech often forms the timberline. Overall, the beech forests are the forests of the south, and as well as fulfilling the important role of protection against the agents of erosion, they provide several natural habitats, recreation and amenity.

All the New Zealand beech species are evergreen and the canopy of mature forest is fairly dense. There are fewer storeys than in the rain forest, fewer small trees, shrubs and ferns, and a more open woodland type of habitat is evident. The terrain, however, is as varied as anywhere in the temperate world: from the flat productive plains of Canterbury and the rolling hills of Southland to the lush bush-lined valleys of central Westland and the steep and remote beech-covered mountains of the south-west. The diversity of forest habitats in the South Island is possibly unparalleled in so small an area.

Natural vegetation and topography are not the only factors that contribute to such a variety of habitats. There is a marked change in climate from west to east. The valleys west of the divide have a high annual rainfall spread throughout the year. By comparison the eastern valleys are much drier but have greater temperature extremes. The influence of climate and terrain upon the forests also affects the creatures that inhabit them.

Even before the Maori arrived in New Zealand there was open grazing country separating the great forests. Here roamed the moa and other flightless birds. But the Maori brought the kuri (dog) and the kiore (rat) and the men from Hawaiiki hunted the moa. When Europeans began to colonise these shores the land, which left to itself might have maintained the delicate birdlife-vegetation balance, was never the same again. Fire and axe, opossums and wallabies from Australia, deer from Scotland, cats, rabbits, mustelids and rodents all took their toll.

Opossums

From 1837 to 1875 some 15 known releases of opossums were made throughout New Zealand; in the North Island the most notable ones were in the Auckland, Wanganui

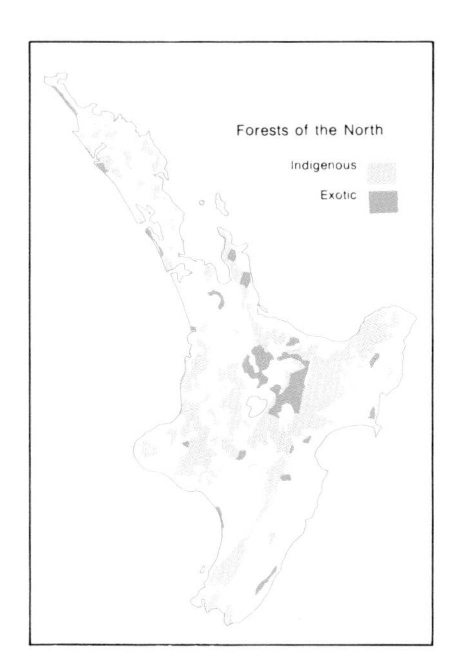

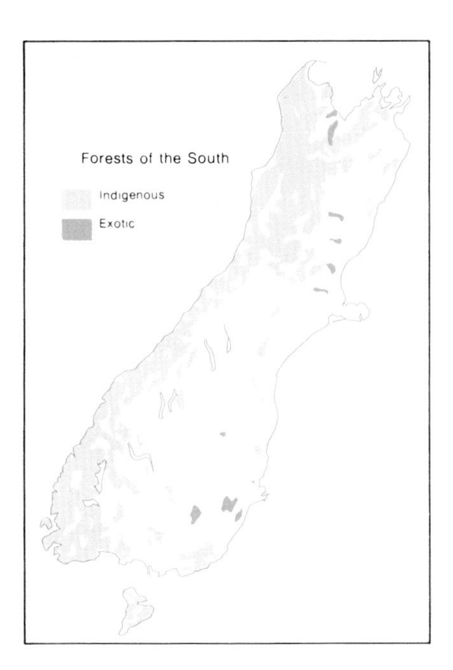

Exotic forests are not as extensive in the South Island as in the North but are equally important to timber-related industries, house building and export.

New Zealand indigenous forest.

and Wairarapa districts; in the South Island the main introductions were in Southland and Canterbury. The reasons behind the introduction were many: sport, keeping as pets, Australian nostalgia, and a little later the promise of a lucrative fur industry.

Nowadays, regardless of climate, opossums can be found in varying density in all types of indigenous forest, from sea level to the upper tree line. They are also found in established exotic forests, shelter belts, orchards and even urban gardens. Feeding primarily on leaves of trees and shrubs, with some fruits and flowers, the brush-tailed opossum causes considerable damage to forests throughout the country. In pine plantations, opossums eat the tender tips of young trees and have caused havoc with eucalypts, poplars and willows which have been planted to stabilise eroding slopes and river banks.

Heavy populations occur in both main islands and Stewart Island; in the North Island the most favourable habitat types are the forests of the southern part of the island, especially around Mt Egmont, the Ruahines and southern Wairarapa; and the Ureweras and South Auckland in the top half. South Island habitats include central Westland, Karamea-Buller district and the north coast. Populations are especially high on forest margins and in remnant forests surrounded by grassland. This habitat differs from dominant forest in that increased food resources are available in the adjoining grasslands.

Like other marsupials the newborn opossum is born at a very early state – only 17-18 days after mating. Cases of twins are rare, though some females may give birth while the older (weaned) young rides on the back. Just prior to the birth the mother adopts a sitting position. The young is ejected free of foetal membranes and immediately turns to face the direction of the pouch. Little more than an embryo, pink, hairless and blind, the baby then crawls through the fur by alternate overarm movements of the forelimbs. On reaching the pouch it becomes firmly attached to one of the two teats. The whole journey takes about 10-15 minutes.

After 10 weeks attachment is not obligatory, although the length of pouch life is not clear cut. The young opossum gradually spends more time out of the pouch, riding bareback from about 20 weeks and may return to the pouch as late as 25 weeks. While generally weaned between 25-35 weeks some young may occasionally remain with the mother until over a year old.

The brush-tailed opossum occurs in two colour types – black and grey – which probably evolved as a protection against predators. The grey is less visible in open forest and scrub where the darker form is better concealed in dense forest.

Bats

Two small bats, each about the size of a mouse, are New Zealand's only known native land mammals. The short-tailed bat is unique to New Zealand appearing to have no close relatives anywhere else in the world, whereas the long-tailed bat is one of 17 species of the Australasian genus *Chalinolobus* with members in Australia, New Caledonia and South Africa. The short-tailed bat is the rarer of the two and although widespread in the last century, the North Island concentrations are now mainly limited to the Urewera district, Rotorua and the northern Bay of Plenty. Long-tailed bats have

Easily distinguished from the long-tailed bat by its large ears and smaller tail, the short-tailed bat is equally at home looking for food on the ground, in the branches or in the air.

The delicate, parchment-like wings of the long-tailed bat measure only 25-28 cm from tip to tip.

been recorded in the same districts as well as Wanganui, the Waikato and the Raukumaras. Both are to be found in the forests of the South Island, the long-tailed being the more common. The largest populations are in the Nelson district, northern Westland, and Stewart Island where a rare subspecies of the short-tailed bat is precariously defying the attentions of the kiore.

Both species are bush-dwelling animals usually seen only at dusk as they flit over clearings or rivers in pursuit of airborne insects, capturing and chewing moths and mayflies with sharp, specially adapted teeth which can inflict a painful bite. Their only enemies seem to be moreporks and the occasional stoat or rat. Short-tailed bats are equally at home looking for food on the ground or on the wing. They have a unique method of folding and protecting their delicate wings which enable them to be more active on the ground and in trees than almost any other species of bat.

While both are usually thick-furred and brownish grey the long-tailed bat may sometimes be almost black, and as its name implies differs in shape from the short-tailed. The latter also has larger ears and is more robust in the limbs. However, due to their erratic flight it is very difficult to distinguish them on the wing and identification is almost impossible unless captured.

Long-tailed bats hibernate during winter in hollow trees or caves whereas the endemic species does not; thus any bat seen flying at dusk or later during the winter months will probably be the short-tailed.

Wild Pig

Captain James Cook has been credited with the introduction of pigs to New Zealand when in 1773 he released what were probably English domestic stock at Queen Charlotte Sound and at Cape Kidnappers. Ever since then the pigs that have been released on these shores have interbred with feral pigs of many varieties to the point where establishing their origins has become impossible. It is, however, safe to assume that the true wild pig was never introduced to this country.

Pigs are still plentiful in forested districts of the North Island, but except for isolated pockets are scarce in the South Island. The largest concentrations are in the Taranaki-Wanganui region and the forested areas between Lake Taupo and East Cape.

Their size and colour can vary considerably, a mature pig sometimes attaining a shoulder height of one metre. Depending on habitat and availability of food the same specimens can weigh up to 180 kg and despite their ungainliness and bulk are extremely agile and fast moving. Protruding from each side of the lower jaw are razor sharp tusks which enable both boars and sows to tear up the ground while searching for grubs and fern roots. Pigs are omnivorous and will feed on roots, berries, or on animal carcass with equal relish. They are also gregarious animals and it is not uncommon to see mobs of pigs comprising all ages. Continual rooting by family mobs tends to modify the vegetation of an area, an 'invasion' resulting in bush or pasture very quickly taking on the appearance of a ploughed field.

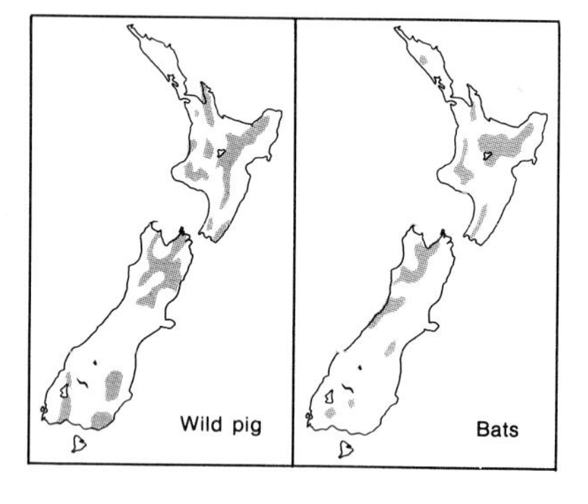

Sharp tusks protruding from each side of the lower jaw enable wild pigs to tear up the ground in search of food. Considerable damage to the forest ground cover is the result.

Rats

Rats and mice have long since spread to most corners of New Zealand. The most recent of the introduced rodents, the ship rat is the most common and widespread of the three rat species in New Zealand – particularly in bush or forested areas. Different colour forms of this one rat are found throughout the country, in rural and suburban districts, from the coast to the high country. They are essentially arboreal and are excellent climbers. Of all rodents, they have had the most drastic effect on New Zealand bird life and must take much of the blame for bringing a number of species to a dangerously low level. It is ironic that their main predator, the stoat, is truly carnivorous, eating anything it can catch, and that includes birds also.

The common colour form in North Island forests is the brown-backed, white-bellied variety. In the South Island most common is the brown-backed and grey-bellied form, the black form making up 30 per cent of the population. It is thought the black form survives better in cool wet climates than the two brown forms and less well in hot dry areas. Rats are notoriously prolific breeders and research has shown that they can breed continuously as long as enough winter food is available.

Norway rats are confined mainly to offshore islands and the kiore's only mainland habitat is Fiordland. It is probable that this distribution pattern reflects the presence or absence of their natural predators – stoats. Where stoats are present, Norway rats are effectively controlled and survive only where abundant food is found close to secure refuge. Kiore, in pre-European times, was widespread, and an important part of the Maori diet. Today it is interesting to note that the decline in their population corresponds with the spread of mice, which suggests that competition from mice has been a major factor in the kiore's decline.

Deer of the Forests

The sika, or Japanese deer, were imported in 1904 and released in the central North Island, and although they have spread slowly through the forests of the Kaimanawa and Kaweka ranges, the main concentration is still in the area surrounding the point of release. They have proved to be very adaptable and in many cases have gained a territorial dominance over their larger cousins, the red deer, as a result of their aggressiveness and intensive foraging.

An elusive, cunning and physically tough member of the deer family, the sika has shown it is well equipped to hold its own against red deer, harsh winters and prospective trophy hunters.

The Javan rusa is the smallest and most localised group of deer in New Zealand having remained close to their original point of liberation in the Galatea foothills on the western

	Sika	Sambar	Rusa
Height (at shoulder)	0.91m	1.20m	1.05m
Weight	80kg	250kg	150kg
Points	8	6	6
Pelage	Chestnut-red, white spots. Black dorsal stripe. White tail and rump patch.	Chocolate-brown. Underparts pale brown.	Reddish-brown. Grey-brown towards winter. Pale chest, neck and underparts.

An identification guide to the deer of the central North Island.

border of the Urewera National Park. They have become largely nocturnal in habit but a careful observer at dusk can sometimes see a rusa stealthily working its way down from a scrub gully to feed on ferns and toitoi along the forest margins. Rusa is distinguished by the pale chest, neck and underparts which are retained throughout the year. The moderately long tail is tufted at the end.

The third forest dweller amongst the North Island's imported deer is the large and impressive sambar. Usually exceeding the size of even the largest red deer and weighing about 250 kg, the sambar is probably the biggest trophy prize of all and hence the numbers are few. In New Zealand the sambar has forsaken the mountain habitat of its Himalayan origins and appears satisfied with its new arboraceous environment. Like the rusa, the sambar has not been unduly perturbed by increased human activity and has adapted by becoming largely nocturnal.

Unlike most other deer, sambar stags do not replace their antlers each year, but do so irregularly every second, third or even fourth year. When fully developed, sambar antlers like those of the rusa comprise six points.

Sika have proved very adaptable and have generally replaced their close cousin, the red deer, in the areas surrounding their original liberation.

In the forests of Fiordland two large species of deer are the main browsers – animals that feed on the leaves of trees as distinct from grazing. The largest of all deer, the moose, was introduced in 1910 when ten animals from Canada were liberated in the Dusky Sound area. Wapiti had already been released near George Sound five years previously, but neither animal has spread very far.

The moose failed to colonise for several reasons, not least being the unsuitability of the habitat in which it was introduced. The southern part of Fiordland consists of deep and precipitous glacial valleys densely clothed in forest and drenched by a massive annual rainfall of some 7,500 mm. This inhospitable habitat has evidently proved too much for the moose and it is doubtful if more than 20 animals now survive. That

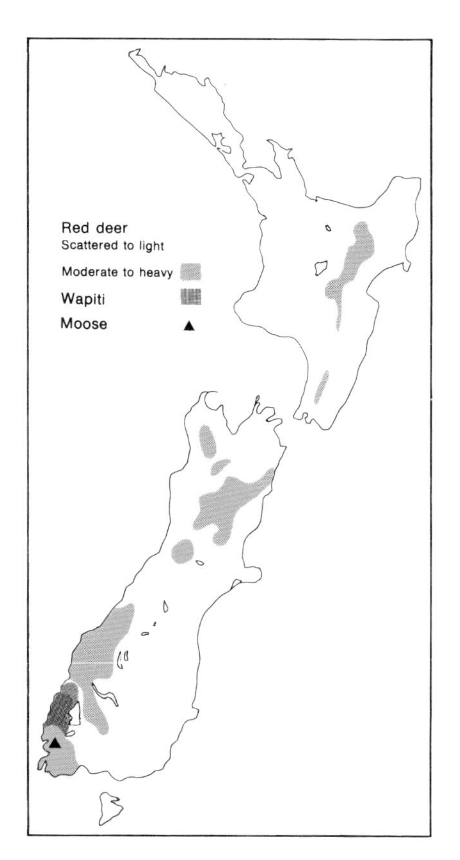

The majestic bull wapiti is the largest round-horned deer in the world.

first small herd also had to compete with a vigorous coloniser already well-established – red deer.

A large, ungainly creature, the moose stands nearly 2 m at the shoulder, appearing to be disproportionate to its short body length, and its weight of 500 kg plus seems to be concentrated around its head and shoulders. The head itself is long with a large flat pendulous nose enabling the animal to completely submerge its head while searching for underwater plants. The most distinguishing feature is the pair of hugh palmated antlers of the male, which on mature Alaskan specimens exceed 2 m in spread.

Because the few moose still struggling to survive are probably nowhere near their optimum size, the wapiti bull would take the title of the largest wild animal in New Zealand. A large, impressive deer, the wapiti has become a prized target for generations of hunters. After 70 years of acclimatisation the wapiti have slowly spread over an area of 3 million ha but natural boundaries have restricted their expansion further afield. Competition with red deer for available browse, as with the moose, has had an adverse effect on population growth.

Belonging to the same genus as red deer, wapiti have long been known to interbreed and fears are held that the extinction of the pure wapiti herd is only a matter of time with hybridisation eventually absorbing the pure strains and the population becoming, in Fiordland at least, a wapiti-cross deer.

Standing about 1.5 m high at the shoulder and weighing up to 365 kg, the mature wapiti bull is considerably more bulky than the red deer stag. However, at a distance the two look very similar, especially in summer when the pelage differs only in the prominent creamy-coloured rump patch which extends upwards over the base of the wapiti's tail. As winter approaches the general colour lightens to a pale straw although the head, mane and legs remain darker. The calves, born during November and December, are richly coloured and spotted like red deer fawns.

Wallabies

Six species of wallaby were introduced to New Zealand from Australia in the last century and all have become established with varying degrees of success. The tammar, dama or scrub wallaby is the only one to have shown a preference for the bush, especially in the Rotorua district and on Kawau Island. In Rotorua they have caused many problems by eating young pine trees and crops. Some of the wallabies removed from Kawau Island have been returned to their homeland where the species has now become rare. Nocturnal feeders, tammar make networks of well-defined tracks through the bush to feeding areas, sometimes foraging in groups.

The Stoat

Of the three mustelids – stoat, weasel and ferret – that have been introduced into New Zealand, only the stoat may be regarded as common and widely distributed. The stoat is the middle-sized one, measuring from 34-40 cm from nose to tail, and its long tail has a conspicuous tuft of black fur at the end. It is brown above and white or yellowish below. Not confined to forest or farmland, stoats are found in a wide range of habitats which also include city gardens and mountainsides as high as the vegetation limit. While no mustelids exist on Stewart, Kapiti, Great and Little Barrier Islands or the Chathams, they have been known to swim to some offshore islands, including Resolution and others in Fiordland and the Marlborough Sounds.

Stoats, like weasels and ferrets, live almost entirely on meat, eating vegetable matter only by accident. Their diet will consist of the smaller creatures of field and forest – rats, mice, birds and insects and even some of the larger ones – rabbits and hares.

Reptiles

Two families of lizards are represented in New Zealand – the geckos and the skinks. The main

Stoats are found countrywide in a wide range of habitats, from farmlands and forests to city gardens and high country.

distinguishing feature is that geckos have a loose-fitting skin covered with small scales, while skinks have a firm, smooth and shiny skin. Geckos are usually nocturnal and they are able to make a variety of sounds from muted chirps to loud barks. Skinks are most active in the daytime, enjoy basking in the sun and are silent and fast-movers. Both periodically slough their skins, and both are able to shed a portion of their tail when escaping from predators.

Most New Zealand geckos have a distinct preference for habitats in or near the bush or forest, and a remarkable ability for climbing hard, smooth surfaces. Some, but not all, have prehensile tails. Of the three genera present in New Zealand, one is represented by three species in the North Island, one is widespread over both main islands, and the third consists of six species confined to the South Island.

The Pacific gecko is found in a wide variety of habitats over most of the North Island. It is nocturnal, sheltering during the day in scrub, forest debris or under loose bark and stones. Only slightly smaller than the forest gecko, they are not easy to tell apart.

The Auckland green tree gecko and the Wellington green gecko are subspecies. The latter is distinctively larger and more heavy-bodied with a bluish-green dorsal colour, the males having blue flanks. When disturbed it can be aggressive and vocal.

The Auckland green tree gecko is undoubtedly the most handsome of the North Island geckos. It is found in manuka and other low shrubs where its vivid shade of green, often with yellow or white markings, provides excellent camouflage. The tongue is black and the inside of its mouth a deep and brilliant blue. In calm and sunny conditions this

New Zealand's second largest gecko, the forest gecko. Note the loose-fitting scaly-looking skin which is characteristic of all geckos.

The Auckland green tree gecko in the process of sloughing. First the outer layers of skin split behind the head and part way down the back. The animal then climbs out leaving the shed skin completely inside-out and often quite whole. Below: *Portrait*

gecko will climb to the top of its favourite tree to bask in the sun and catch numerous flying insects. The colder months of the year are spent in semi-hibernation, sheltering under large stones or beneath the bark on the lower part of its host plant.

The Northland green gecko is a close relative confined to the very north of the North Island. It is very similar in appearance, slightly larger and displaying a red tongue when threatened. Sometimes entirely yellow specimens of both these species may be found.

The common gecko is the most pervasive of all New Zealand's geckos, occurring widely throughout the country.

As widespread and nearly as ubiquitous as the common gecko, the forest gecko is the most attractive of all our geckos and as its name implies, lives mainly in the forests of both islands although it may sometimes be encountered above the bushline in the South Island. The colour and the pattern of this species, like so many of the others, is extremely variable. The pattern especially can range from simple streaks to intricate designs and the colour, usually shades of brown, may include combinations of gold, grey, black, green and white. This forest dweller is also able to change the intensity of these colours according to its background, the effect being enhanced by remaining absolutely motionless when disturbed, and thus becoming virtually invisible. It is found in open forest frequently half-hidden under stones or logs or the loose bark of trees. Though nocturnal, this reasonably large species (18-19 cm) often emerges during the day to bask in the sun.

Possibly the most handsome of the New Zealand geckos are the six species of the genus *Heteropholis* which are confined to the northern parts of the South Island. All are diurnal, all are tree-dwellers and all are green. They vary in size and shape and their patterns range from almost plain to the most exquisite arrangements of lines, spots, diamonds and blotches. The starred gecko which has a limited range from the Nelson Lakes to Golden Bay is one of the most striking of all our lizards. Some specimens bearing brown shapes on a green background, others the reverse, typify the variability in colour and pattern in this one species.

Known only from specimens found in the Lewis Pass-Reefton area, the Lewis Pass green gecko is a large and most beautiful animal. Irregular shapes in different shades of green and paler spots outlined in black give a mossy appearance, making it very difficult to spot in its preferred habitat of shrubs and bushes close to beech forest. A strong aggressive tendency is evident in the male, its 'threat' being accompanied by a series of loud barks. This species has the loudest voice of any New Zealand gecko.

Geckos (and skinks) shed the outer layers of their skin at intervals. In summer most species will slough once every two months. The animal appears pale and dull as the top layers become loose and detached from the underlying ones. The whole process is generally completed after about two hours. Despite the brilliance of some species the slough will always be a light grey because the skin pigments lie much deeper and are not lost in shedding.

Frogs

Three frogs have been introduced to New Zealand: two very similar green tree frogs, one which failed to establish itself and one other which is found in close proximity to streams and marshes, and the whist-

The starred gecko is diurnal and lives in areas of scrub or open bush in the Nelson district.

A typical Auckland green tree gecko habitat. The manuka flowers produce a nectar which attracts the nectar-feeding tree gecko.

The Jewelled Gecko

The native shrubs, small trees and pine forests of two separate regions of the South Island provide suitable habitats for a gecko with more variations in colour and pattern than any other within the genus. The jewelled gecko is found in Canterbury and Otago with 200 km between each centre of population. Not surprisingly they differ in colour to such an extent that they may eventually be regarded as separate subspecies. (Robb 1980)

The southern form – the Otago Peninsula jewelled gecko – is predominantly dark green with an assortment of yellow or white stripes or patches. The Banks Peninsula form in the Canterbury region is more often a shade of brown or grey-brown, multicoloured markings adorning its flanks. Like many of the arboreal geckos this species has a prehensile tail which is used to help them move through the branches, almost as if they had a fifth leg. Species which use the tail in this way tend to lose it less often.

Two forms of the jewelled gecko: above, *from Otago and* right, *from Banks Peninsula.*

These handsome lizards blend so inconspicuously with their backgrounds that finding them is not easy – especially amongst the foliage. Most specimens are likely to be seen during the late autumn or early winter and almost invariably they are pregnant females about to give birth.

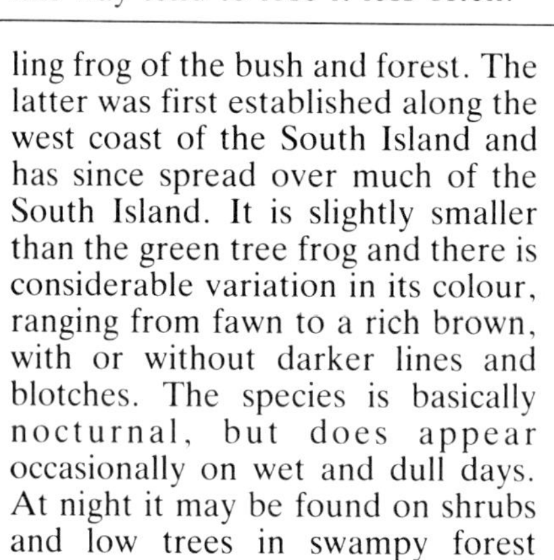

ling frog of the bush and forest. The latter was first established along the west coast of the South Island and has since spread over much of the South Island. It is slightly smaller than the green tree frog and there is considerable variation in its colour, ranging from fawn to a rich brown, with or without darker lines and blotches. The species is basically nocturnal, but does appear occasionally on wet and dull days. At night it may be found on shrubs and low trees in swampy forest independent of water except when breeding. Slow moving water in ditches or streams is frequented during spawning, each female releasing hundreds of tadpoles during the season. The name comes from the shrill, chirping call of the male frog.

Frogs Without Tadpoles

Three rare frogs in the genus *Leiopelma* are New Zealand's only native amphibians, and two of them are restricted to relatively small areas of North Island forest. The habitat of the third is confined to two islands in Cook Strait. All three are very small, ranging from 28 mm to 38 mm, and all are generally brownish, the smallest, Archey's frog, being slightly mottled with green. The territories of both the North Island frogs overlap in the high mountain forests of the Coromandel Range – well away from open water. In fact, Hochstetter's frog is the only one of the three that may be encountered in the vicinity of water, and is

The whistling frog prefers to live in areas which provide cover during the day, cool temperatures and adequate moisture.

Hochstetter's frog was the first native frog to be discovered. Its colour varies from brown to green-brown and is the sturdiest of the three native species.

the most widespread of the species.

The primitive characteristics which distinguish the native frogs from other species are somewhat subtle but the most fascinating aspect is their life history. All frogs need moisture and die if exposed to drying winds or sun. The native frogs, however, have evolved so that the laying of spawn in water has been by-passed. All development takes place in a gelatinous capsule derived from the egg originally laid by the female, and the young emerge as tiny froglets with tails. The tail is lost after a few weeks. Archey's frog has thus developed with no webbing at all between the toes and Hochstetter's frog has only minimal webbing. The native frogs are fully protected and should not be disturbed.

Creatures of the Forest Floor

Amongst the layers of slowly decaying vegetation on the forest floor, whether it be a carpet of pine needles or a tangle of rotting fern, dwell countless millions of small creatures, the hoppers, millipedes, slaters, earthworms, some snails, slugs, and larvae of many insects, which feed on the decomposing

A mature nymph of the common green shield bug.

litter. This litter layer is inhabited almost exclusively by these invertebrates, all part of the food chain which includes larger animals and the forest itself. There is an extremely complex relationship between plants and insects. While some forest insects feed on a wide variety of plants, others are restricted to very few, perhaps even a single species. However, the relationship between different species of insect is even more complicated. One species inevitably depends for its survival on the activities of others, either singly or collectively. As logs and branches decay, different insects invade at different times, one specialised species preparing the way for the next; beetles scavenge for dead or helpless insects; larvae of countless kinds depend on rotting vegetable debris; predatory flies and slugs and snails rely on moisture to survive. Many are remarkably adaptable and are consequently found under many conditions. Some, such as earthworms eat the dead leaves directly; others depend on fungi or bacterial decay to break down the rough tissue of vegetable matter. Many are so tiny that they simply are not visible to the naked eye.

Snails, and their relatives the slugs feed on nearly all organic debris by rasping it with long, strap-shaped tongues, or radulae, covered by rows of small teeth. The majority of the herbivorous snails are tiny – some as small as a pinhead – but the native slugs may exceed 10 cm. Some snails, such as the large *Paryphanta* and *Powelliphanta* species are carnivorous, feeding on worms, insect larvae and occasionally fleshy fungi. Adult specimens of several species of *Rhytida* also prey on the juveniles of their own species!

New Zealand has over 200 land snail species catalogued with probably many more awaiting

A native land-snail of the Paryphanta *genus and a typical forest litter habitat.*

discovery. Although well distributed throughout the country the number of species decreases as one moves south. In the forests of the Auckland region, for example, it is possible to collect thirty species in a day; twenty in the vicinity of Wellington.

The present distribution of large land snails in New Zealand shows that conditions were once far more favourable than they are now, and subfossil remains point to a greater number of species with extensive ranges. The North Island habitats have suffered more than those of the South Island but rats, mice and hedgehogs continue to pose a threat. Wild pigs are also a significant cause of diminishing populations. Perhaps the largest number of giant species is nowadays to be found in the forests of Marlborough and Nelson where both lowland and highland species exist in fairly high numbers.

Land slugs are air-breathing snails which have all but lost their shells (just a few calcareous granules remain buried beneath the lung). Apart from the common garden slug introduced from Europe, New Zealand has 23 known species of native veined slugs, all of which are typically marked by a leaf-vein-like pattern of groves and markings on their upper surface.

The species represented in or near the forest litter, is a fungivorous feeder which spends its periods of inactivity under logs.

Millipedes also feed on leaves, mosses and fungi and decomposing plant material. Their legs are short and well adapted for pushing the almost cylindrical body through loose forest debris. Though often locally abundant, millipedes are extremely secretive, seen only when logs or stones are turned over and soil disturbed.

The ancestors of New Zealand's veined slugs were probably among the original molluscan colonisers of Gondwanaland.

It is not surprising that only very few of New Zealand's 600 species are distinguishable at first sight. The smallest are a mere 2-3 mm long and live under dry bark; some North Island species, however, attain lengths of 10 cm or more.

Unlike millipedes, centipedes are predators and have few enemies. Some reach 10 cm in length, but generally are much smaller and very slender. The largest New Zealand species *(Cormocephalus rubriceps)*, however, may attain a length of 20 cm north of Auckland, and the poison carried in its pincers can produce a reaction in sensitive people.

Centipedes and millipedes both belong to the group Myriapoda, meaning 'many-legged', although they rarely possess as many as 100 and never as many as 1000 respectively! The main obvious differences are the projecting pincers of the centipede and just one pair of legs for each body segment, whereas the millipede lacks the pincers and only the first few body rings have one pair of legs, the remainder two pairs.

Also present in the leaf mould are the worms, amphipods and isopods, silverfish – and springtails. Springtails are tiny, wingless, jumping insects, some so minute that tens of thousands can live in a square metre of damp soil. They are immensely important in the breakdown of forest debris. *Holacanthella paucispinosa* is a species found not only in leaf litter, but also in old logs powdered through decay. It is the largest known springtail growing to a length of 14 mm and a breadth of 5 mm, and is covered with needles rather like a miniature hedgehog.

Arachnids

Most important among the forest litter predators are the arachnids; the eight-legged creatures which include spiders, mites, harvestmen and pseudo-scorpions.

There are probably about 2000 species of spiders in New Zealand, only half of which have been described scientifically, and an

A trapdoor spider of the tunnelweb variety takes up an aggressive stance.

amazing variety of extraordinary habits and behaviour is evident. Spiders differ from other arachnids in several ways. The two portions of the body, the cephalothorax and the abdomen, are joined by a narrow waist. Poison fangs are used to subdue their prey and the second pair of leg-like appendages (the pedipalps) are used to transfer the sperm to the female during mating. Unique to the animal kingdom is the production of silk from the spinnarets at the base of the abdomen.

Well represented in New Zealand are the trap-door spiders and their relatives. Though often large they are not poisonous to man. Some, like the large, black tunnelweb spider *(Porrhothele antipodiana)*, are occasionally found inside the home trapped in bath or sink, but most of our hundred or so species are seldom seen. All live on or near the ground but some construct a loose silken tunnel with a mouth opening on the side of a loose stone or crevice. The true trapdoor spiders excavate a burrow in the ground and usually seal the mouth with a hinged lid cunningly camouflaged with forest debris. The tunnels may be 25 cm or more in length, the spiders remaining there throughout their lives extending them as they grow. All trapdoor spiders are heavy-bodied and rather squat, brown or black. The female is always stouter than the male.

The tree-trunk trap-door spiders, rarely more than 1 cm long, construct the whole nest complete with lid on the trunks of trees or rock faces. Orbweb spiders usually construct their webs during the hours of darkness and these are often plainly visible in the light of early morning by the dew which adorns them. Despite the apparent complexity and the unrivalled efficiency and strength, the whole procedure involved in a web's construction may be completed in half-an-hour. The tailed forest spider is at home in shaded places and is commonly found in forests. The abdomen is usually yellow-brown, but sometimes red. The orbweb of this spider always has one segment missing and it is here that the row of egg-sacs is strung.

The harvestmen vary considerably, some attired in grotesque knobs and spines. While most spiders capture their prey alive, the harvestmen are scavengers, consuming almost anything from dead animals to berries and decaying vegetation. They detect their prey with extremely long legs and then hold it with pincer-like fangs while the food is slowly ground into small pieces. Unlike spiders they have no poison to subdue their prey and no silk to entangle it.

Some mites are predators too, not only on other mites but on a wide variety of small life. Amongst these the centipede and the ancient peripatus – the link between insects and segmented worms. The peripatus is shy and retiring and seldom seen. Mites emerge at night to hunt for prey, living and dead. Unable to control water loss from their bodies they seek shelter during the day in damp surroundings, showing a remarkable ability to squeeze into the smallest spaces.

The female tailed forest spider grows to about 2 cm including its long flexible tail-like abdomen: the male is minute rarely exceeding 2 mm.

This grotesque harvestman, Algidia chiltoni, *rests motionless during the daytime under stones or logs but moves at night to feed.*

Abundant among insect predators are the beetles, of which there are more different species than any other animal group. They are to be found everywhere, the forest litter being a favourable habitat especially for weevils, stag beetles and longhorns which all require dead wood. The litter also affords shelter to countless soft-bodied larvae, that satisfy the predacious rove and ground beetles. The largest are the carabid or ground beetles. Usually black with a metallic sheen, these forest-floor carnivores are commonly found in and under decaying logs. At night both the adult beetles and their voracious larvae hunt the smaller insects on which they feed.

Weevils are beetles with snouts. The longest New Zealand beetle is the giraffe weevil whose snout and head together make up half its length (up to 7 cm). The teeth at the extreme tip of the snout must move by remote control. Seeming to sense when a tree is about to die the adults congregate on the trunk and the larvae bore into the solid wood before it starts to decay.

Male and female giraffe weevils, New Zealand's longest beetles may attain a length of 8 cm.

Stag Beetles

New Zealand has at least 25 species of this group, but not all are endowed with the aggressive-looking horned mandibles which give them their popular name. The abdomen is comparatively small in this group, and they seldom take to the wing. The larvae live in rotten wood, often in the galleries of primary wood-borers; adults are usually found under old logs in the bush.

The Helms' stag beetle is typical of those with very prominent mandibles which may sometimes be used for fighting between males, but have a more likely function for burrowing. The female carries a similar but smaller pair. Helms' stag beetles are found only in the bush of the west and south of the South Island and Stewart Island.

Herbivorous insects depend not only on the decaying litter but also the leaves and flowers of living plants. Much damage is done to native trees by a host of these animals which include cicadas, aphids, the caterpillars of moths and butterflies, stick insects and wetas. But we have to remember that plants and their insect associates have evolved together, and disaster occurs only when the natural balance is upset, usually by man's interference.

Wetas

Largest and probably best known of the forest insects are the wetas. This fearsome looking relative of the cricket, can be divided into two distinct groups: cave wetas, and tree and ground wetas. The former are fairly common throughout the North and South Islands, Stewart Island and even on some of the subantarctic islands. Wetas are primarily vegetarian, feeding on wood and leaves, usually at night, though they may sometimes be seen on green leaves in bright sunshine. They have very long slender antennae and massive hind legs which can inflict a painful kick when disturbed. The distinctive rasping sound, produced by the hind legs striking the ridges on the abdomen, can be heard in the bush at night. Although rather heavy they are surprisingly agile and can jump long distances amongst the branches or leap along the ground. In fact, wetas have been referred to as 'invertebrate mice'.

Cave wetas are particularly well represented in New Zealand with over 50 known species. Despite their name they are not confined solely to caves. They are secretive creatures, seeking out dark, humid habitats often in the bush, under stones, logs or in hollow trees. Cave dwelling species are modified for a cave environment: legs and antennae are usually greatly elongated, colour is duller and the eyes are smaller. The large *Gymnoplectron* prefers the twilight zone near cave entrances where it forms large congregations emerging independently after dark, to feed in the bush.

Cave wetas are primarily scavengers, living on a mixed diet of animal and plant material. The bush species, probably the most familiar, are often associated with ferns. Their noiselessness and dull mottled appearance make them blend easily with the background. They have no

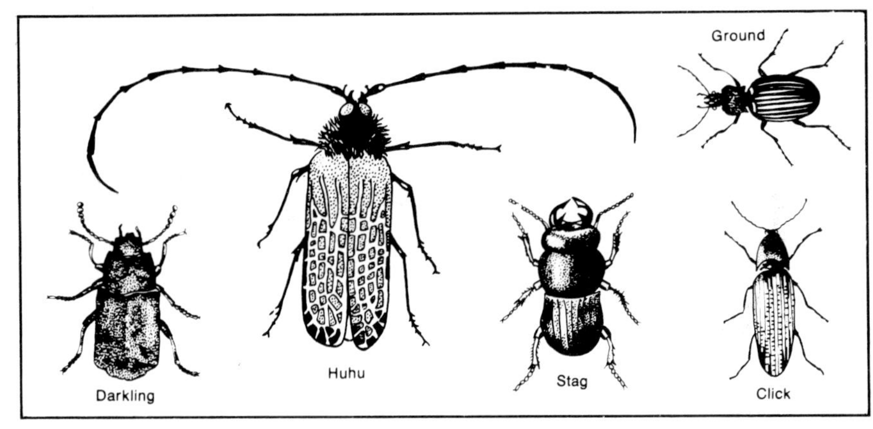

There are more species of beetles than any other insect group and they are to be found almost everywhere. Forest leaf litter is often a very favourable habitat for many of New Zealand's 4000 native species, supplying them with a variety of dead leaves, twigs and fungi.

Above: *Cave wetas are generally smaller, a little less fearsome in appearance and have evolved much longer antennae.* Below: *Tree and ground wetas have a characteristic mottled brown colour, well-developed jaws, large bodies and sturdy, sharply spined hind legs.*

wings, small jaws and their enormously long hind legs give them great jumping powers. The adult female can be distinguished from the male by its long, curved ovipositor projecting from its rear end.

Herbivorous insects are not confined to the forest floor. The beech canopy supports a richer leaf eating fauna than the kauri-podocarp-hardwood associations and apart from some of the wetas, the chief defoliators are caterpillars, stick insects and grasshoppers. Insects which eat plants often betray their presence by modifying a part of the host they inhabit. Tied leaves, drooping buds, a spray of wilted foliage and various deformations (galls) reveal the activity of a plant-eater. Chewed leaves signify the attentions of caterpillars or beetles. Larvae of some small flies and weevils feed between the leaf surfaces forming 'mines'. Each year the kowhai loses most of its leaves to the caterpillars of the kowhai moth. Yet the following year it blooms again, seemingly little the worse.

New Zealand has a large group of moths whose caterpillars tie leaves together and feed on the inner surfaces. Others mimic their host plant very closely. Among them is the caterpillar of the zebra moth. Fully grown larvae rest motionless appearing just like a twig. Others curl up to resemble a beech leaf, complete with crinkly margin picked out in red. One looper caterpillar imitates the knobbly stem of the whipcord hebe with green and yellow blotches arranged to the best advantage.

Mantids and Stick-insects

The praying-mantis and the stick-insect are arboreal creatures occurring mainly in the warmer parts of the world. Both are very difficult to see when stationary relying for protection on their resemblance to their surroundings.

In New Zealand there are only two species of praying-mantis and about twenty species of stick-

insects. One of the mantid species is a real mystery. First described about 100 years ago it seemingly disappeared altogether until rediscovered in Auckland in 1978. The mantis is carnivorous, its food consisting wholly of other insects, mainly flies. It patiently stalks its victim and when within striking distance sweeps its front legs, armed with formidable spines, to catch its prey. The female mantis is similar to the male but has a stouter abdomen. After mating she lays her eggs in a soft, frothy case which is attached to a branch or twig and hardens on exposure to the air. Mantids are renowned for the fact that the female will sometimes eat the male during mating.

Stick-insects are vegetarian, feeding on a limited range of plant species. The largest local species is a spiny, grey-brown creature with a body length of up to 15 cm. Grey, green or brown, these interestingly-shaped creatures can conceal themselves very effectively by remaining absolutely motionless, thus resembling a leaf or twig. Their main enemies are birds, wasps, rodents and lizards. Many species of stick insect can reproduce without the eggs being fertilised. Where males occur, fertilisation may take place but is not essential. This phenomenon is known as parthenogenesis, and it is thought that fertilised eggs give rise to male progeny, unfertilised eggs to female. Males are invariably smaller than females, and during mating will remain mounted on the female's back for days. The eggs hatch into tiny replicas of the adult and there is no grub stage.

Above: *A praying-mantis in characteristic pose.*

Two of New Zealand's 19 species of stick insect: one a smooth brown creature, the other spiky and green.

Bees

Other insects, such as bees and true flies (Diptera), are important to the well-being of the forest. They are the ones which assist in the pollination of flowering plants. Several types of bees have been introduced, the functions of which are mainly agricultural and it is the small native bees (family Colletidae), together with the honey bee, which are the most useful.

There are about 40 species of native bees, all smaller than the honey bee. Found throughout New

Native bees are important pollinators of many native plants, particularly tea-tree, pohutukawa and rata.

Zealand they can be very numerous in some areas, swarming around tea tree, rata, pohutukawa and other native trees.

The native bees are fairly primitive when compared with honey and bumble bees. They have neither a queen nor workers, but are divided only into males and females. Besides laying the eggs the females do all the work in the nest; the males do nothing but mate with the young females. Most species make nests in tunnels that they excavate in the ground; those that do not, build their nests in hollow branches. In both cases the female bee forms a small cavity at the end of the tunnel where it deposits a little lump of pollen and nectar. One egg is laid on the food and the cavity is sealed. The mother bee never sees her young so that colonies like those of honey bees are never developed. A female bee makes about twenty such 'nests' in six weeks or so, then dies. Native bees are usually black and are often overlooked because of their resemblance to flies. A weak sting is present in the females but even when disturbed they have not been known to attack.

Flies

One of the most successful orders of insects is the Diptera – the flies. The most easily recognised characteristic of flies is that they have only one pair of wings, whereas almost all other winged insects have two pairs. New Zealand has over 2,000 species spread throughout every conceivable habitat. Some, especially those that associate with humans – houseflies, midges, mosquitoes and sandflies – are well known, but the vast majority are not obvious. It is peculiar that the best known are the least liked, the 'nuisance' flies, that conjure up ideas of filth and disease. But there are also very attractive flies and beneficial flies. Craneflies, for example, comprise hundreds of species, which inhabit moist forest areas and some are beautifully coloured with oranges and greens. Hoverflies, often called flower flies, are large and attractive and mimic other insects. One such member is the drone fly which not only looks similar to the drone honey bee, but also searches for nectar. Soldier flies include several endemic species ranging from 6 mm to 15 mm long. Some of them are brightly coloured.

Glow-Worms

Unlike European glow-worms, which are beetles, the New Zealand glow-worm is the larval stage of a luminous species of gnat. The Waitomo Caves are world famous for their glow-worm grottoes, where thousands are visible in the darkness, each emitting a bluish-green light from its tail end. But glow-worms are not confined to caves nor to the North Island.

The larva of the glow worm builds itself a tubular nest of mucus and silk in which it suspends itself.

Larvae of common hoverflies live amidst aphid colonies which they devour as food.

Wherever the habitat is damp and shady glow-worms are likely to be seen, often forming impressive displays with their twinkling lights. Likely places in the forests are stream banks, crevices in the ground or the dark and sheltered bush itself.

The larva suspends itself from its canopy by a series of fine silk threads, each bearing at regular intervals a small droplet of mucus. These 'fishing lines' can vary in length from 1 cm to 5 cm, and are used by the larva to catch its prey. Attracted by the luminescence the midges which breed in the water and the mudbanks below them, fly upwards and become ensnared in the fishing lines. The larva then hauls up the line on which the midge is attached and consumes its prey in its nest. The hungrier a glow-worm is, the more brightly it glows, the light organ being formed from the excretory organ which is an extension of the gut.

The life cycle of the gnat is passed in four stages: eggs, larva, pupa and adult. When the adults emerge from the pupa they are still glowing, and the female is usually fertilised immediately by one of the waiting males. Neither adult lives longer than about four days.

Birds of the Forests

Before the arrival of Man, New Zealand was indisputably a land of birds. Free from predatory animals, birds thrived and were certainly the most dominant class of vertebrates. Man's inroads into the forest-clad interior, the clearing and burning of large tracts of land and the browsing and grazing of mammals that accompanied the early settlers, inevitably reduced the country's avifauna. Between the Polynesian and European settlements, the moa, a pelican, two flightless geese, a crow, two harriers and several ducks and rails all became extinct.

Birds that once occurred in countless thousands have dwindled with the depletion of the forests and the influx of predatory animals since man's arrival. In cultivated areas competition with exotic species is so keen that native birds are in the minority. Isolated lakes, primeval bush and dense virgin forests are now the last refuges of many of our native birds, especially the vulnerable flightless species, such as the takahe, weka, kiwi and kakapo.

And yet New Zealand is still a land of birds: not in its wealth of species but in their uniqueness. Due to its long isolation, there are among this countrys exciting bird fauna, several families that occur nowhere else.

The Puriri Moth

There is one endemic family of moths, nearly all members of which inhabit the forests. Included in this family is the large green puriri or ghost moth, the female sometimes attaining a wing span of up to 15 cm.

The caterpillars of this moth bore into native trees, notably puriri, titoki, putaputaweta, ngaio and wineberry, and feed on the bark and underlying tissue, concealed by a silken tent constructed around the tunnel entrance. The duration of the larval stage is not accurately known, but is believed to exceed three years and may possibly be longer than five.

The adult, our largest moth, is nocturnal in habit and has a strong attraction to light. The female produces about 2000 eggs which she seems to scatter indiscriminately. When first laid they are yellow but soon turn black and take about two weeks to hatch. Newly hatched larvae search for decaying wood on the forest floor where they complete their first growth stages. After a year or so they leave their dead wood haven in search of a suitable tree in which to spend the rest of their larval lives. Then begins the construction of a tunnel which serves as both home and larder, gradually extended as required. When fully fed the larva enters the pupal stage which may last up to six months, during which the pupa remains remarkably active. Tiny spikes around each segment act as levers to assist its movement within the tunnel. Emergence occurs between August and February, peaking in October and November. In contrast to the larval stage, adults live only a few days at most, just long enough to mate and lay eggs.

A North Island brown kiwi in search of food. All kiwis feed on grubs, insects and worms supplemented by leaves, berries and seeds.

There is still forest in parts of the North Island (especially on some of its offshore islands) which offers a glimpse of how primordial New Zealand must have looked. These are the places where the whimsical kiwi and the shy wattle birds continue to defy civilisation's progress; where the flightless weka patrols the undergrowth and honey-eating birds sing in the canopy.

Kiwi

Represented by several species the kiwi is not found outside New Zealand.

All kiwis are robust, strong-legged birds with feathers that resemble a shaggy coat of hair. They have no tail, their reduced wings are almost invisible, they are nocturnal and their eggs are disproportionately large. When foraging at night the bird uses its long and very sensitive bill to probe deep into the litter and soil for worms, grubs and beetles. Tapering holes are a typical sign of the bird's recent presence, but most obvious of all is its tremulous and piercing call 'ki-wi' which inspired the Maori to give it its famous name.

Only one species, the common, or brown kiwi, inhabits the northern half of the North Island and though preferring the indigenous rain forest may also be found in scrub and tussock. The brown kiwi, like its South Island relatives, has poor vision despite its nocturnal habits, but its senses of smell and hearing are thought to be very highly developed. All species have the same general dietary habits: earthworms, woodlice, slugs and snails, spiders and centipedes, and a variety of insects. The egg of the brown kiwi averages one-fifth of the female's weight; a greater proportion than is known in any other bird.

Three subspecies of the brown kiwi are found respectively in the three main islands. The South Island brown kiwi is similar but larger than its North Island cousin, and its colouring is somewhat lighter. Populations are to be found in scattered western districts. The South Island has an additional species, the great spotted kiwi, and a third species, the little spotted kiwi is now known only on Kapiti Island and D'Urville Island.

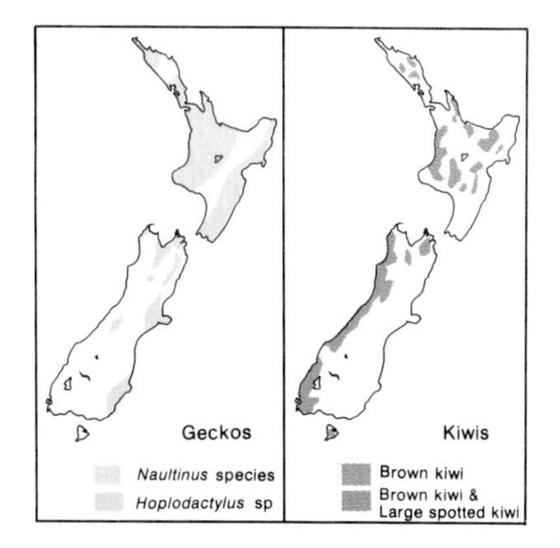

The great spotted kiwi is the largest species, weighing up to 3.7 kg, about the size of a well-fed domestic fowl. Confined to the forests west of the Main Divide, it is more common in the north than in the south.

The Weka

A flightless member of the rail family the weka is well known for its inquisitiveness and will often stride boldly into the open, help itself to any item of a campers' food or equipment that takes its fancy and strut back to the concealment of the bush. Of the four endemic subspecies three were originally South Island birds, the rarest being the buff weka which now is probably restricted to the Chatham Islands. The more common bird of the South is present throughout the West Coast forests and farmlands, and the

Great spotted kiwis occur only in the South Island in western districts. Distinguished from the little spotted kiwi by the chestnut tinge on its back.

third is confined to Stewart Island.

Two colour phases occur in the South Island weka – the typical chestnut plumage is found throughout the bird's range, but in the far south a black form is not uncommon. Hybrids in the overlapping areas are of intermediate colouring.

The North Island subspecies of the weka now survives naturally only in the Gisborne area although the race has been successfully re-established in Northland and Coromandel. Being a terrestrial bird the weka possesses only rudimentary wings, but is well equipped with large, strong legs and feet.

Takahe

Another flightless rail, the takahe, is one of the rarest birds in the world. Once widely distributed, it is now confined to alpine tussock and forest margins in the Murchison Mountains region of Fiordland. This larger, high-country relative of the pukeko is declining in its sole remaining habitat due mainly to the spread of red deer competing for available food.

Kakapo

Even rarer than the takahe is an owl-like, ground-dwelling parrot – the kakapo. This most peculiar bird is a true parrot; soft, textured plumage, facial discs and its nocturnal habits are, however, all characteristic of the owls. It cannot fly but climbs well using its strong claws and bill, then glides back to earth again. The wings are used mainly to balance itself when running or climbing. Its owl-like face is adorned with 'whiskers' growing from the base of the bill which, like those of the kiwi, appear to be organs of touch.

Though clumsy in appearance the kakapo is beautifully patterned with light green above, yellowish-green below and irregularly flecked and barred with black and brown. It blends perfectly with the vegetation of its natural habitat, hiding by day and using well-marked paths at night in search of favourite grasses on which to feed.

Kakapos prefer the mossy beech forest up to the tree line and the alpine meadows above, but also live in fern thickets, mountain flax and other shrubs. Areas of avalanche debris are particularly favoured, where shelter may be found among huge rocks, in the cracks and crevices they form and amongst the vegetation they protect.

Predominantly vegetarian, feeding on berries, leaves, fern fronds and roots, kakapos have also been seen eating lizards. A peculiar method of feeding is characteristic of the bird: it chews blades of tussock grass, rolling it into pellets and sucking out the juice, before leaving it dangling from the stem fibrous and desiccated.

Until 1977 the chances of the kakapo surviving seemed remote. The population was estimated at a mere twenty birds and they were all presumed to be males. However, the discovery of several individuals on Stewart Island in February of that year has renewed hope that

Four subspecies of weka are recognised. This, the South Island, or western weka, is common west of the Main Divide.

Above: *Sinbad Valley in Fiordland, the main kakapo habitat.* Below: *A male kakapo in his courtship 'bowl'.*

extinction has been deferred. A feature of kakapo country is the shallow, bowl-shaped depressions once thought to be dust bowls. It is now known that these 'bowls' and the system of tracks connecting them are used by the males as display grounds during the breeding season. From the 'bowls' which act as sounding boards, sited against tree roots or the lower side of large rocks, male kakapo make resonant booming calls which carry across the valleys at night. The calls, and the remarkable displays which accompany them are thought to attract females but so far no female has ever been seen to respond.

Two parrots of more typical behaviour also inhabit the forests of the south: the kea, better known as an audacious scavenger of the high country, and the kaka.

The Berry-Eaters

Amongst the larger bush birds are the berry-eaters, and the most handsome of these is undoubtedly the native pigeon. Its iridescent green and purple feathers were much prized by the Maori for cloak decoration and its flesh was regarded as a delicacy. Still widespread, the pigeon remains a fairly common resident of the indigenous forests, and is now protected by law, though many are shot illegally. This unwary bird feeds chiefly on the fruits of native trees and is an important distributor of seeds of many species. Perhaps because it is heavily built, it will indulge in spectacular stalls and swoops when flying in the open, across a valley, for instance.

About the same size as the pigeon but less frequently seen is the brown parrot or kaka. While fairly common throughout the main islands and the larger offshore islands, the North Island subspecies survives in greatly reduced numbers in the kauri forests of North Auckland and the indigenous forests of the Island's central ranges. The South Island kaka is larger than the North Island subspecies, has more grey on the head and is tinged green on back and wings. Its feathers were much sought by the early Maori, especially the vivid red of the underwings.

Large areas of podocarp forest, are preferred but isolated patches of bush are sometimes visited. Kaka diet consists of a wide assortment of

The plump and handsome native pigeon (below) *rests in a totara tree, and* (right), *the vivid South Island kaka, the larger and brighter of the two subspecies.*

fruits such as miro and matai berries, nectar from flowers of kowhai and rata, leaves, bark and insect grubs dug out of the ground. Its sensitive tongue is used with the beak half open to tap the bark of a likely tree in search of grubs. A neat incision is then made where the 'sonar' has given a positive lead. When extracting nectar, the bird often uses one foot to hold the flower.

The Honey-Eaters

The New Zealand representatives of this predominantly Australian group of very specialised brush-tongued birds consist of two birds whose melodious songs give pleasure to all who hear them, and one other bird rarely seen or heard at all. The best known is the tui. Its beautiful, liquid metallic notes and the beat of its wings can often be heard in native bush or as it flies over suburban gardens. The iridescent blue-green plumage – though appearing black at a distance – and the prominent white throat tufts, are matched in splendour by its amazing aerial displays. Especially at dusk, the tui may be seen performing fantastic aerobatics in mid-flight, sometimes appearing to stall before dropping fast and vertical.

Only two-thirds the size of the tui, the bellbird is difficult to see in the bush, its greenish hue affording effective camouflage. Though its song may resemble that of the tui its habits differ in that it thrives in all types of forest where, particularly before sunrise the air will be filled with its delightful bell-like chorus. During the last century it became extinct north of Auckland, probably through disease. It is now slowly returning, but has not yet reappeared in the Waitakere Ranges near Auckland City.

The bellbird includes a high proportion of insects in its diet which it extracts from crevices in the bark with its distinctly curved bill. However, it will stray considerable distances from its breeding area, especially in winter, to locate its favourite sources of nectar – kowhai, puriri, rewarewa, fuchsia, rata and pohutukawa.

The stitchbird, almost threatened with extinction, is restricted to Little Barrier Island off North Auckland's east coast. Once common over much of the North Island, this colourful honey-eater proved to be extremely susceptible to change, but appears to be thriving on its island sanctuary.

Parakeets

Of the four species of parakeet native to the New Zealand region only the red-crowned and the slightly smaller yellow-crowned parakeets are North Island residents; the former being uncommon on the mainland but plentiful on off-lying islands, the latter being generally widespread. It is not understood properly why the yellow-crowned parakeet has maintained its population when the red-crowned parakeet has almost disappeared from the mainland. Perhaps an introduced avian disease has taken its toll of the latter. The yellow-crowned parakeet generally keeps to the upper tiers of the foliage where they chatter while feeding on berries, buds and seeds. They are more often heard than seen, except when in flight when their vivid colours are unmistakable as they flash past.

The four species are representatives of a sub-family of parrots common in Australia. The native species are medium-sized birds ranging in length from 20 cm to 33 cm, all bright green with blue on the wings and differing only in the colour detail on their heads.

The orange-fronted parakeet, a smaller species, is rare and confined to the forested slopes of mountain ranges in the Nelson district and Lake Sumner. When viewed from a distance, it is rather difficult to distinguish from the yellow-crowned parakeet. The fourth parakeet is Forbes' parakeet, an endangered species now restricted to the Chatham Island.

The introduced Eastern rosella, twice as large as the parakeet, was brought to New Zealand from Australia originally as a cage bird. The rosella now occurs wild in many bush-clad areas north of the Waikato and in the Upper Hutt Valley where its strikingly brilliant plumage and long tail make it easily distinguishable. Its partiality for fruit make it disliked by orchardists.

The Small Birds of the Forest

The rifleman and the silvereye are two of New Zealand's smallest birds; both are abundant where there is plenty of tree cover and both are mainly green in colour. The rifleman is the most common of the New Zealand wrens, inhabiting indigenous forest south of Auckland province and on Little and Great Barrier Islands. Its food consists almost entirely of insects taken from the bark of trees and the forest floor.

Unlike the rifleman, the silvereye belongs to a large family of about 85 species ranging throughout the Southern Hemisphere. A conspicuous white ring around the eye and its longer tail feathers avoid confusion with the rifleman. It is much more common, often visiting suburban gardens in small flocks searching for the various fruits and nectar of popular garden plants.

New Zealand's nine species of flycatehers and warblers include several of the country's best known and loved bush birds. Some of them have a number of regional variations making 22 subspecies in all. The fantail is the friendliest and most unmistakable: never still for more than a few seconds, long tail continually raised, fanned and lowered, twisting and turning in flight in search of tiny airborne insects. It is one of the few native birds that has adapted to changed conditions following European settlement. Minor plumage differences distinguish the three races of the North and South Islands and the Chathams, the 'black' form being common in the South Island, rare in

The metallic sheen of the tui's plumage shown to its best advantage.

The yellow-crowned parakeet appears to be gradually increasing in numbers throughout both main islands, especially in the larger forests.

Four of the smaller birds of bush and forest: (clockwise from top left) the familiar pied fantail; the trusty South Island robin; the inquisitive North Island tomtit and the ubiquitous silvereye.

the North Island and absent from the Chathams.

The pied tomtit, widely distributed throughout the country, finds its food on the trunks, branches and the floor of native and exotic forests, clinging sideways to the bark as it scans the litter for the movement of insects. There are five geographic races all strongly territorial in the male; the pied tit of the North Island, the yellow-breasted tit of the South Island, the Chatham Island tit, the Auckland Island tit, and the black tit of the Snares.

One of the most vocal of the bush birds is the grey warbler, whose gentle trill may be heard throughout most of the year. Not confined to the bush, it is also a common bird of scrubland and even populated areas. The larger Chatham Island warbler, with different plumage, is classed as a separate species, but behaves similarly to the mainland species.

Restricted to the larger areas of bush and exotic forest in the southern two-thirds of the North Island, the whitehead is a bird of the tree-tops, though it sometimes seeks its food of insects and seeds on the ground. Its cup-shaped nest, bound together with cobwebs, is usually made in manuka or tree ferns and is a favoured site for the eggs of the long-tailed cuckoo.

The North Island robin was once common but its population has now retreated to the Ureweras and the forested ranges from the Mamakus south to the Tararuas in sparse distribution. The South Island race, however, is well represented on some Cook Strait islands and is irregularly distributed from Marlborough down the western side of the South Island to the southern lakes and Foveaux Strait, but are absent throughout almost all of Canterbury east of the Alps. An exception is a remnant of lowland forest near Kaikoura which is well known for its variety of birdlife, the robin being its most numerous native inhabitant. Even more numerous is the Stewart Island race that inhabits some offshore islands which are free from predators.

The North Island subspecies is a little larger than a sparrow, almost black above with creamy white underneath. The male is streaked white on head and throat, and the legs of all subspecies are much longer than those of the tits or grey warblers.

Robins construct their nest of twigs, mosses, leaves and grass woven together with spiders webs, and lined with feathers and hair. Their food consists of worms, grubs and insects; the male feeding his mate while she is nesting. But most captivating of all is the song of this delightful little bird. As fluent as a canary with the power of a blackbird, the recital may continue unchecked for 20 minutes, filling the air with an amazing range of notes.

The Morepork

The morepork is primarily a bush-dwelling bird, but it will be heard wherever suitable trees or shrubs give sufficient cover. It is the most common of the three owls in New Zealand and the only one in the North Island. In the South Island beech forests the predatory morepork is becoming scarce. With the introduction of stoats the morepork's main source of food – birds and native rats – is being depleted. A nocturnal bird, it remains concealed during the daytime, usually in a tree or rock hollow, emerging at night to feed on insects, small birds, mice and lizards.

Both the Maori name, 'ruru', and 'morepork' are names derived from its call, but the listener's imagination could probably devise other descriptions when the bird is heard close-by on an otherwise silent night. They also make a screech, possibly when hunting. Only the female incubates the two or three eggs which take a month to hatch. It is a further five weeks before the downy chicks leave the nest.

The Kokako

This shy and attractive member of New Zealand's wattlebird family, once widely distributed, is now becoming increasingly uncommon. There are two subspecies, one inhabiting the more inaccessible forests across the central portion of the North Island, Northland and Great Barrier, the other barely surviving in a few isolated areas in the South Island and Stewart Island. However their numbers are dwindling and extinction, certainly of the orange-wattled, South Island subspecies, may be imminent.

The blue-wattled kokako is about the same size as a magpie, dark bluish-grey in colour with a broad band of black extending from the base of the bill to around the eye. Not a great deal is known about this bird; it is particularly shy and nests high up in the forest canopy. It is a poor flier, having a clumsy, bounding action over the ground or

Above: *A morepork returning to its nest.* Below: *The North Island kokako is still widely distributed if not as numerically strong as it once was. This female and chicks were photographed in the Coromandel Range.*

through the trees. Mainly vegetarian, much of the birds' time is spent on outer branches eating young leaves and berries, including the fruits of bushlawyer, supplejack, pigeonwood and the native fuchsia. Some variation of diet is provided by small insects found in lichens and mosses covering the trunks of old trees.

According to Maori folklore, the kokako was the only bird of the forest which brought Maui a drink when the mythical hero returned thirsty and exhausted from a trip to the sun. As a reward Maui stretched the kokako's legs to make them long and slender, thus enabling it to run and climb more nimbly through the trees.

The Brown Creeper

Widely distributed in the South Island the brown creeper is completely unknown in the North Island. It is smaller than a sparrow, and is found in all types of forest, including exotic plantations. They are gregarious birds associating together in small flocks often accompanied by silvereyes and yellowheads. Their call is a rapid succession of metallic trills as they diligently hunt for insects among the tree tops.

The nest is always well-concealed 2 m or more above the forest floor. A neat cup, wider at the base than at the rim, is woven from moss and grass, bound with spiders webs and lined with feathers. Two to four eggs blotched with purplish-brown and denser at the larger end, are laid in mid-summer. Unhappily for the brown creeper, the nest often plays host to the long-tailed cuckoo which destroys the brood in those nests it selects for its own eggs.

Cuckoos

Two cuckoos return each summer to breed in New Zealand after spending the winter in the Pacific islands. The shining cuckoo is only the size of a sparrow and heard more often than it is seen. The long-tailed cuckoo is a large, hawk-like bird, whose scream is heard most often at night. Both cuckoos are widely distributed in forested areas during the summer months, the shining cuckoo favouring willows in close proximity to urban areas, the long-tailed cuckoo purposefully spreading to exotic plantations.

The plumage of the long-tailed cuckoo is a striking combination of bars, spots and streaks on a dark brown upper-surface and pale buff under-surface. Cuckoos do not build nests but lay their eggs in the nests of other birds, mainly of the grey warbler and brown-creeper, ejecting the eggs of it's host in the process. Because only a single egg is laid in a nest, it is not known how many eggs each female will lay in a season.

Both species feed mainly on insects but the long-tailed cuckoo will also include lizards and small birds in its diet, as well as the eggs of its host.

Predatory birds are few in New Zealand. The kea, the morepork and the weka, are also known to hunt small animals but the only true birds of prey are the harrier hawk and the falcon. The hawk relies considerably on carrion but the New Zealand falcon has been known to take tuis, kakas and pigeons and even rabbits and hares. It is a fearless hunter, swift on the wing and fierce in combat. In the forest it hunts by flying silently through the trees and attacking its prey from below.

Introduced birds have increased throughout the country in all kinds of habitats. The greatest number of species is to be found in open country but exotic forests are also congenial to such birds as blackbirds and thrushes, finches and larks. The chaffinch and greenfinch have penetrated deep into the plantations of both islands. Pheasants and California quail are common and appear to be increasing.

The brown creeper, though fairly common in the South Island, is completely unknown in the North. The nest is generally well-concealed in the forest canopy or in a jumble of bush lawyer and other vines. Opposite: *Long-tailed cuckoos arrive mostly in October and breed from Auckland to Stewart Island. Their distribution, however, depends largely on the presence of suitable foster parents.*

Urewera National Park

The Urewera National Park is one of the few places in New Zealand that has such a prolific variety of forest cover. It ranges from luxuriant kohekohe, large tawa and rimu, to pockets of mountain beech and wind-shorn scrub. Soil, rainfall and altitude dictate the variety.

As one ascends from the valley floor, species appear, assume abundance and then fade from sight as other hardier associates take their place. From any high peak one has a view of the almost continuous forest cover over the 400,000 ha of the park. Deep river gorges of the Whakatane, Waimana and Waioeka, hot springs at Waiohau and in the upper reaches of the Waiau River, cascading waterfalls and quiet lakes add character to the park.

The Urewera is often thought to have few birds and only a short list of different species. Unless one is sitting quietly by a lake or clearing, or tramping through the bush this appears to be so. Birds such as the North Island kiwi, weka, and morepork, are frequently heard but seldom seen. Pigeons, tuis, tomtits, riflemen, bellbirds, grey warblers and fantails may be seen feasting on nectar and berries or darting from branch to branch looking for food. The fantail seems to be the most common, accompanying trampers or flitting around on the edge of a clearing. The rare North Island robin and kokako continue to increase within the park.

Forest clearings are favoured by hawks, pipits and kingfishers as they watch the ground for prey. On the lakes and rivers are found dabchicks, shags, black-backed gulls, white-faced heron, and native ducks fishing and swimming. The introduced mallard, however, is proving harmful because of its interbreeding with the grey duck, consequently reducing the pure population.

Both New Zealand's native land mammals, two species of bat, are reported to live in the Urewera. Strictly protected, they live in colonies in hollow trees, caves and disused mine shafts. Being generally nocturnal they are rarely seen in flight, flying mainly at night in search of aerial insects.

Two or three geckos and at least one skink are also tenants of the Park. Most likely to be seen amongst

Urewera bush with Lake Waikaremoana in the background.

The Asian rusa deer is the smallest and most localised deer in New Zealand being restricted to western areas of the Urewera National Park.

Open forest, like this glade of silver birch, is a favourite haunt of many introduced birds.

the geckos is the Auckland green tree gecko in various shades of green. Unlike most other geckos this species is unable to change its colour according to its surroundings. Instead, its prehensile tail is extremely versatile enabling the animal to swing from one branch to another.

The observer's eye is often caught by the ornate skink's metallic sheen glistening in the sunlight. It may be glimpsed on dry banks or bare ground close to low vegetation or rocky outcrops to which it retreats rapidly when threatened.

Greenfinches, the largest of the finches, thrive wherever pines are grown. In autumn some flocks resort to coastal paddocks and feed among banded dotterels.

Introduced animals include pigs, deer, goats, opossums and a few wild cattle that have escaped from domesticity. Sambar deer are fairly well established, though not frequently seen on the forest margins; sika have spread from their point of liberation into the southern Urewera forests and rusa are well established on the western margins of the Ikawhenua Range. The most aggressive colonisers, here as elsewhere, have been the red deer which now occupy most of the Park. Gross modification and significant damage has been wrought by red deer and opossums, their browsing and foraging preventing regeneration of native trees. Goats from adjoining farmland are also responsible for much damage, and being the least selective of browsers, they have seriously retarded vegetation in many areas.

Hanmer Springs State Forest Park

Nowhere in New Zealand is there such a wide variety of exotic trees growing in forest proportions. Covering an area of nearly 10,000 ha is a wealth of shape and colour which includes larch, pine and fir among the conifers and poplar, silver birch and mountain ash among the deciduous trees. Altogether about 40 different species are represented. The area is abundant in birdlife with many pleasant walks that allow keen and casual observers alike to see both native and introduced birds in profusion. Patience is the only requirement; most birds are inquisitive enough to make themselves visible when noise and disturbance is minimal. Birds likely to be encountered quite quickly are the grey warbler, bellbird, fantail and tomtit (maybe an occasional brown creeper, robin, rifleman or weka) and, of course, the blackbird, chaffinch, thrush, greenfinch and hedge sparrow.

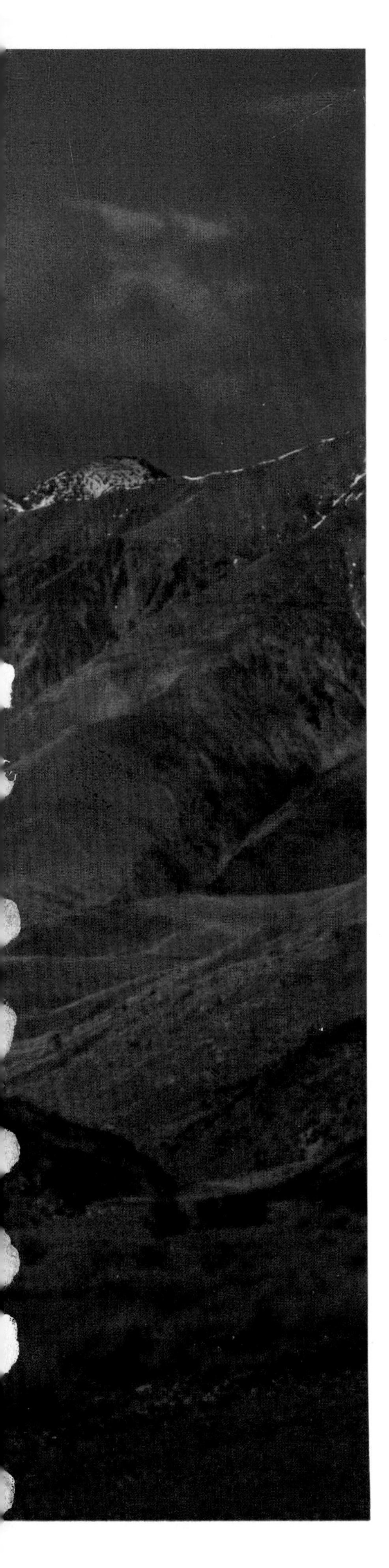

HIGH COUNTRY

Animals which are to establish successfully in an alpine or subalpine habitat need different adaptations from those that live in the more hospitable lowlands.

All terrestrial environments are protected in various degrees by the atmosphere. This insulating blanket of air prevents plants and animals from being damaged by the sun's radiation during the day, and retains some of the accumulated warmth at night, so that the earth's surface does not become unbearably cold like the moon.

The higher the mountain, the less dense becomes the atmosphere and the more its protection is lost. Temperature decreases progressively, high altitude winds increase in speed, frost and snow break down the rocks, and constant rain washes away what little soil exists.

In a short vertical distance a mountain will have as many life zones as there are in thousands of miles of latitude. A fit and able person in New Zealand can, in a matter of hours, climb from luxuriant lowland forest into barren rock and snow.

Forest creeps up most of our mountain slopes, but there is a limit above which trees cannot survive. This limit is called the timberline and it varies according to factors such as latitude and rainfall. Above the timberline, trees become few, stunted, and give way to shrubs, grasses, mosses and lichens – the fellfields. Similarly, the snowline – the line above which snow is almost permanent – varies with local geography and latitude. For example, snow accumulates at a much higher level on Mt Ruapheu than on Mt Tutoko; both are peaks of some 2,800 m. In the North Island, only three mountains rise above the level of potential grasslands, and only one is high enough to support glaciers. Yet in the majestic alps of the South Island there are hundreds of peaks whose crests extend above the heights of perennial snow.

The animals that live on the mountain slopes face severe problems: food is scarce and the climate is cold. Though few large creatures thrive above 1000 m, there is a wealth of other organisms both introduced and native.

The Thar and the Chamois

The largest of New Zealand's alpine mammals are the prized thar and chamois. Both were introduced to the Mt Cook region – two liberations of each – between 1904 and 1914. Both are distantly related to the common goat and both are very much at home on the near vertical cliffs high above the timberline.

The thar, slightly the larger of the two, standing a metre at the shoulder, and weighing nearly a 100 kg, is a very impressive animal in its black winter coat. The bull thar looks even larger because of its dense layer of body hair around the neck and shoulders. Female thar lack the ruff and rarely attain half the weight of the males. Both sexes have short, powerful horns, on which the rings are an indication of age. Few have more than fifteen rings.

In the thar's 400,000 ha breeding range of the Southern Alps their favourite choice of habitat is likely to be between 1300 m and 1800 m on a fifty degree, north or east facing grassy slope. Steeper bluffs would be close by where the animals can retreat if disturbed. However, as the keen trophy hunter and the commercial helicopters infringe their preferred habitat, it appears as though the thar is gradually moving to higher, more precipitious terrain around the 2000 m mark.

If the thar is, perhaps, the most accomplished rock climber in the animal kingdom, then the chamois must be one of the most nimble and graceful. The chamois is the most abundant and widespread of our introduced hoofed animals after the red deer and feral goat. Similar in size to domestic goats, they have longer legs, a more erect neck and pointed ears. A fully grown adult male stands about 80 cm at the shoulder and weighs 40 kg. The does are slighter in build and seldom exceed 35 kg. In summer their coats are short-haired and fawn-coloured with a prominent dorsal stripe. The thick winter coat is uniformly dark brown to black, and the colour

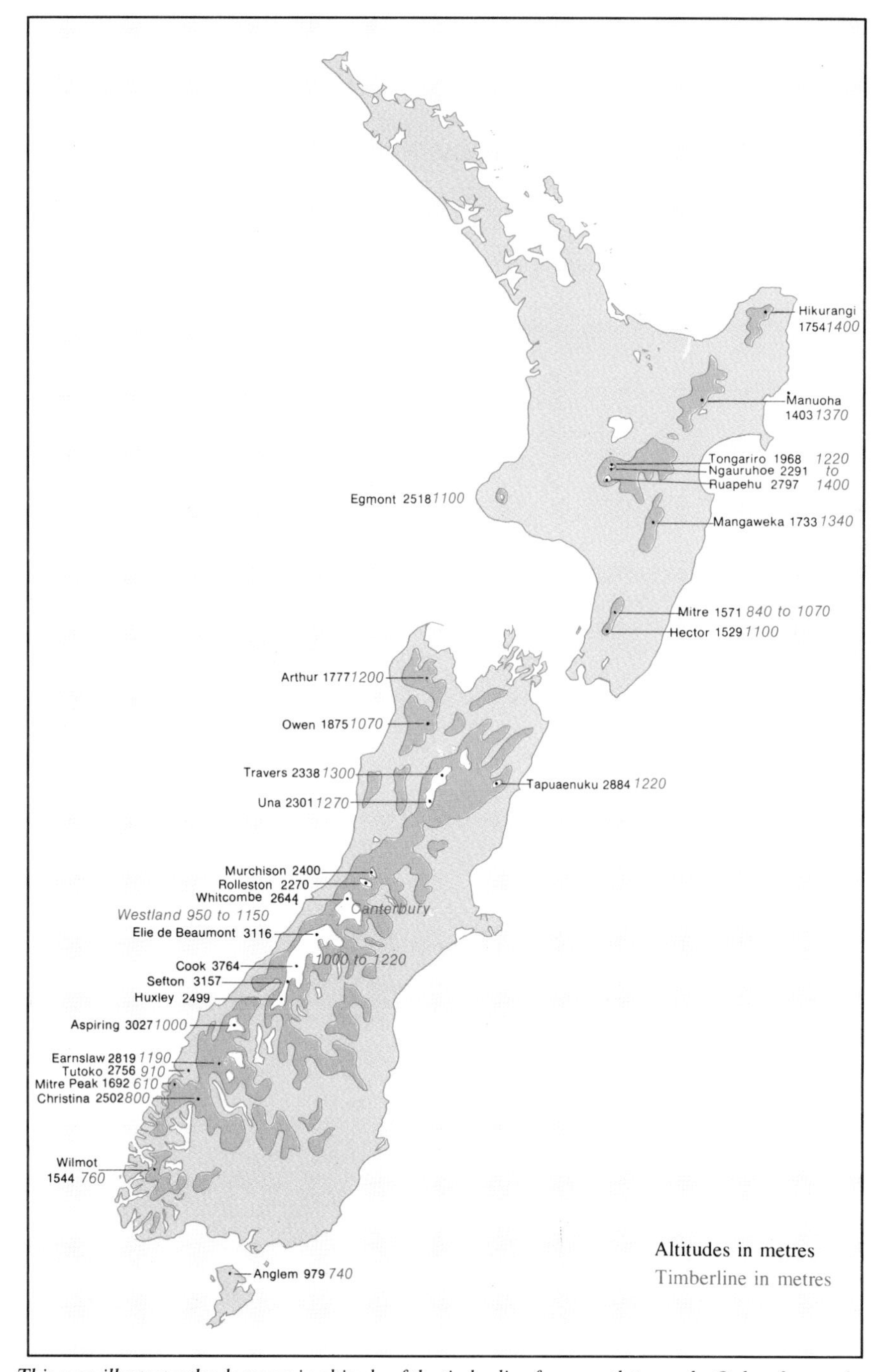

This map illustrates the decrease in altitude of the timberline from north to south. Only a few peaks exceed the 1500 m. timberline in the East Cape region, but in southern Fiordland many peaks of similar altitude soar far above the 850 m timberline.

pattern of the head is distinctive: mainly creamy-white with a dark band beginning on either side of the nose, surrounding the eyes, and ending at the base of the ears. The most characteristic feature of the chamois are the horns, which grow vertically from the brow and then curve back to form sharp hooks.

Chamois, unlike members of the deer family, have extraordinarily keen eyesight which they depend upon to a greater extent than their sense of smell. Their agility is legendary and when disturbed they can cover the most difficult terrain with speed, grace and ease. The kids are mobile from birth and can keep pace with their mothers when only a few days old.

Though often found at high altitudes, chamois do not favour snow covered terrain, but prefer scrub and exposed grassy or scree surfaces near the timberline. In bad weather and during the winter months they seek shelter on steep scrub-covered bluffs where snow does not accumulate.

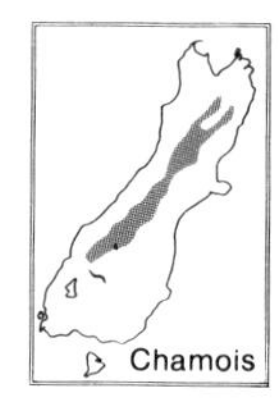

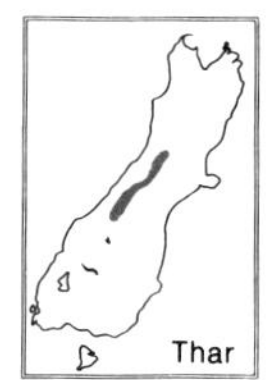

Wapiti and Red Deer

Larger than either the thar or the chamois, but not strictly alpine in their habitat, is the majestic wapiti and its close cousin the red deer. After 70 years of acclimatisation the wapiti has spread over 3 million ha of Fiordland high country forest and subalpine tussock, an area bounded by Sutherland Sound in the north, Doubtful Sound in the south and Lake Te Anau in the east. The Tasman Sea forms its western range where the wapiti also roams down to sea level.

Alpine flora and the forest itself has suffered considerably in the Fiordland National Park because of the browsing of wapiti and red deer. Protection of the latter was lifted in 1930 and wapiti and moose also lost their protection in 1935. All deer species are now fair game for anyone in the National Park, and if the forests and subalpine flora are to survive, wapiti and red deer populations will have to be reduced and controlled.

Although thar are highly skilled rock climbers, they are less adept on steep ice or hard snow. They pick across it very cautiously with none of the confidence they show on rock.

Above: *A young red deer stag in subalpine grassland.* Right: *A wapiti cow with two red deer-wapiti hybrids in the background.*

The red deer is the most common and most widespread of the eight deer species found in New Zealand. Since their original introduction in the Nelson district in 1851, and over a hundred liberations up to 1924, this species has dispersed so rapidly that the density is now fairly even in hilly country from the Bay of Plenty to Southland. Although found in the open country of Marlborough the greatest numbers occur in areas where forest, scrub and subalpine grasslands lie close together.

The absence of predators and an abundant food supply were the main factors governing their rapid increase. Numbers reached such proportions, however, and damage was so extensive that large-scale culling was started in the 1930s and commercial hunting from helicopters in the 1960s. The European market – especially West Germany – took most of our venison exports, but in recent years controlled farming of red deer has assured the long-term export of meat and the more immediate demand of antler velvet in the Far East.

The record length for antlers from a New Zealand red deer stag is 1.28 m, the greatest number of points 40. Any head over 90 cm in length and spread, having 10 or more points, is considered a worthwhile trophy today.

The Hare

The hare is found from rolling paddocks at sea level up to the alpine grasslands close to 2000 m. Introduced in the Canterbury district in 1851, the hare has spread far and wide, establishing itself in the most suitable habitats. Superficially similar to the rabbit, the European hare can be distinguished by its larger size, proportionately longer hind legs and ears, its redder colour and its characteristic loping run. More timid than the rabbit, the hare when disturbed scuttles to the nearest cover or burrow.

Hares are mainly nocturnal, spending most of the day in a grassy 'form' or depression in the tussock. Whereas rabbits are gregarious, the hare is a solitary animal with a territorial range of 300 ha. It is entirely vegetarian eating blue and snow tussock in alpine areas, and grasses and clover on lower agricultural land. In some parts of the country they are regarded as pests because of the damage they do to young trees and also because they compete with domestic stock for grazing.

A curious aspect of reproduction in hares is their ability to conceive a second litter before the first is born – a process called superfoetation.

The Wallabies of the Hunter Hills

The only wallabies to be found in the South Island are the red-necked (or brush) variety in the Hunter Hills of South Canterbury. Of the six species brought into New Zealand this has produced the largest and most extensive population. Their habitat is mainly tussock grasslands under 1000 m with pockets of bush and scrub in the gullies. Being vegetarian, they have proved extremely damaging to crops and pasture, and their numbers have to be periodically controlled.

An adult male red-necked wallaby stands about 75 cm and weighs about 18 kg, sometimes more, making it the largest of the marsupials in the country.

Above: *Similar in size to domestic goats, chamois are very mobile from birth.*

Below: *Hares, like rabbits, usually feed after dusk, and continue to do so intensively until midnight. This male hare is probably on a courting venture.* Right: *The red-becked wallabies of the Hunter Hills comprise the largest wallaby population in New Zealand.*

Geckos

The forests that flank the high country of both major islands provide suitable environments for the two most common geckos of the genus *Hoplodactylus,* the forest gecko and the brown gecko. Both are well distributed and have been recorded up to 1700 m in the South Island. The latter is very variable in colour and pattern, though the basic colours are usually brown, dark grey, blue-grey or olive-grey and black. The tongue and the inside of the mouth are pink.

Like all the other New Zealand geckos this nocturnal species – growing to a length of 12-14 cm – gives birth to live young. Of the 500-600 species of gecko known, only the New Zealand ones are live-bearers.

Cicadas and Grasshoppers

These common but unrelated members of the insect world are well represented above the timberline throughout New Zealand.

Cicadas, whose 'songs' are usually associated with the summer heat are related to aphids and bugs. Sometimes they are wrongly called crickets which make a whistling sound, or even locusts which are silent. It is only the male of the species which sings, to which the female responds by flying or walking towards the singer. Hybridisation is uncommon because each species plays its own tune: some of the smaller species have quite a pleasant song, but the larger cicadas can be somewhat strident.

Twenty or so species of cicada are found in New Zealand. The alpine and subalpine species are members of two genera: *Kikihia* are green, *Maoricicada* are black. *K. subalpina*, although subalpine in its North Island habitats – Tongariro National Park, Mt Egmont, and the Ruahine-Tararua system – is a dweller of the forest canopy in the South Island. It is replaced in the subalpine grasslands of the South Island by the clock cicada, looking similar to its northern cousin but lacking the red spots on its legs. Its song is different too, a monotonous ticking, hence its name.

At Cobb Dam, in north-west Nelson, the Tasman cicada, which is the local subalpine species, is dimorphic. About half the population is green, the other half orange-yellow.

The black cicadas of the genus *Maoricicada* are nearly all alpine. The cassiope cicada of the North Island is a handsome red-black form that is the only black to reach the North Island mountains, occurring on the central volcanoes and the Kaimanawa and Ruahine ranges.

The South Island cassiope is common in Nelson and Marlborough, extending south to the Mt Cook region, where the similar yodelling cicada becomes the prominent species to the east of the Main Divide. This southern species is red-black also, but differs slightly in shape, and with a yodel type of song in place of the cassiope's excited shriek.

Many smaller cicadas are to be found in parts of the South Island, some species adapted to an even higher altitude around 1500 m. As far as is known, nowhere else in the world have cicadas colonised such inhospitable alpine environments. Alpine cicadas in general cannot tolerate the high rainfall west of the Divide. The only exception is the great alpine black cicada which ranges from the Lewis Pass to Fiordland.

The Orthoptera order of insects includes five families represented in New Zealand. They are the crickets, the wetas, the cave wetas, the katydids (long-horned grasshoppers) and the grasshoppers. All are well distributed, though different species prefer different habitats. The grasshoppers are restricted to open country, and all except the well-known migratory locust and one lowland species, are alpine. Of the remaining 13 species, 12 are confined to South Island habitats above 1000 m, some occurring as high as 2400 m.

Despite being only a few kilometres apart, many alpine grasshopper populations are completely isolated by permanent snowfields or deep, precipitous river valleys. No single species ranges throughout the high country and while some prefer tussock others have adapted well to screes or boulders.

On the tussock flats below the Hermitage in the Mt Cook National Park, a large, brown weta is

The rough-scaled gecko is a diurnal species frequently found in scrub and rocky tumbles in the Kaikoura Ranges.

The cassiope cicada is the only one in the genus Maoricicada *– the black cicadas – to reach the mountains of the North Island. It is common in the northern half of the South Island.*

A mountain grasshopper of the Queenstown area.

frequently found when boulders are turned over. Between 1800 m and 2300 m another species in the cave weta group, known as the Mt Cook flea, is likely to emerge in showers when disturbed in rock crevices. There is a large, un-named spider frequenting the same habitat which probably preys upon this weta.

Some of the alpine weevils (snout beetles) are considerably larger than their lowland counterparts. The elegant black and white striped speargrass weevil can attain 3 cm, while an earth-brown species living on *Astelia* plants, is only slightly smaller. They are protected from all except keas and rats by their horny outer covering.

Alpine Moths and Butterflies

Of over 200,000 known species of moths and butterflies, there are in New Zealand about 1500 species of moths distributed in 35 families and a meagre 17 species of butterfly represented in only two families.

To distinguish between moths and butterflies is fairly easy if not always obvious. Generally, butterflies fly by day and moths by night, butterflies are usually brightly coloured, moths often dull. Most butterflies sit with the upper surface of their wings

Kikihia subalpina *is a species which favours the alpine scrub of the North Island high country but lives in the forest canopy in the South Island.*

meeting over their back, while moths rest with their wings fanned out. Closer inspection will show that their antennae differ also: those of the butterfly are club-tipped, while those of the moth are either tapering or plume-like.

The alpine zone in New Zealand occupies about one-fifth of the land area, yet this inhospitable realm plays host to two-fifths of our moth species. Though some are closely related to forest-dwellers of lower altitudes, many are exclusively alpine. As may be expected the South Island has, by far, the greater number of species. A characteristic of alpine moths is the instinctive ability to fly close to the ground in a storm – where the wind speed is less – and seek cover. Lowland species

An alpine weevil of the Anagotus *species on tussock.*

that stray to higher altitudes are helplessly blown away.

Most of these moths are small and inconspicuous; some are very large. Among the larger varieties are the porina moths, several of which are alpine, making their tunnels in the ground and emerging to crop the tussock grasses. Some species are responsible for pasture damage at lower altitudes. New Zealand's largest moth, the ghost or puriri moth, is a member of the same family (Hepialidae).

The Geometridae is a large family easily distinguished in the caterpillar stage. They are sometimes called inch worms or loopers because they move by stretching out the front part of the body and drawing up the rear to meet in a loop. The barred upland looper moth is found only in the high country of the South Island, often in open, exposed situations. It is quite common on the lower slopes of Mt Cook.

Similar in size is the yellow and brown looper moth, found throughout the South Island's alpine regions and on the subalpine scrub of Mt Egmont. Oddly, it is also a lowland species in the vicinity of Invercargill. This species has a ground colour of yellow or orange

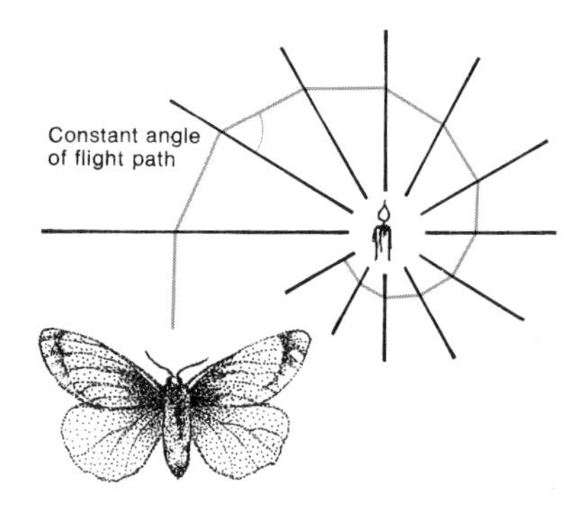

When flying towards an artificial light, a moth spirals inwards in a series of straight lines at a constant angle to the source of the light.

A selection of common moths of the high country.

which makes it more attractive and easier to identify.

Amongst the more common alpine moths four *Notoreas* species are most likely to be seen by the casual observer, but distinguishing between them is not easy. All have forewings of grey and hindwings of orange. Three are residents of the South Island and one, *N. vulcanica,* is peculiar to the central volcanic region of the North Island. Other *Notoreas* species inhabit the South Island south of the Waitaki River.

Butterflies are usually more colourful than moths, and the high country species are no exception. Preferring altitudes in the region of 1500 m, the tussock butterfly is quite conspicuous on the wing but difficult to spot when at rest amongst the tall grasses. The undersides of the wing are streaked in silver enabling this attractive insect to blend successfully with the tussock blades. Butlers ringlet butterfly, which also possesses a pattern of silver marks on the under surfaces of its hind wings, is our rarest endemic species. It has always been found in association with the tussock butterfly, and is similar also in that the female is paler than the male and rarely flies.

Another ringlet butterfly confined to the South Island is the black mountain ringlet which can be seen above 1500 m on rock or shingle screes. It uses the strong updrafts to soar back and forth close to the ground.

Our smallest butterfly is the boulder copper – one of three 'coppers' in New Zealand representing a large family common throughout the world. Though only 2 cm across the wings, the male is one of our most brilliantly coloured butterflies, with a rich purple iridescence on its forewings. It is more common in the South Island than the North, and is often extremely abundant on stony riverbeds or sandy screes up to 1400 m.

Jumping Spiders

Not all spiders build webs to catch prey. Some have become most adept at catching insects without the use of snares, but stalking and capturing their food by stealth. Like most other spiders, jumping spiders have eight eyes but they are arranged differently with a row of four in front – the middle pair being quite large – and two pairs behind, one small, the other large.

Jumping spiders have been found in almost all types of habitat. They have colonised the seashore, the forests, riverbeds and most remarkable of all, the high mountain tops. Some of the most interesting ones live above the snow-line of the South Island mountains. These are large, black-bodied spiders that dwell beneath the stones, and include the white-bearded mountain spider. This distinguished, mottled-grey spider is found above 1500 m in Otago but its camouflage colouring makes it rather difficult to see among the rocks.

Other types of spider also lurk in the mountains. An undescribed species, probably belonging to the Australian genus *Miturga,* has been known to inflict a venomous bite on would-be collectors in the Grampian Mountains of South Canterbury. It is advisable when studying spiders not to handle them with bare hands.

The white-bearded mountain spider, one of the more familiar species of the Southern Alps.

Birds of the High Country

The competition for food and territory that exists between species of the lowlands is not as obvious in the inhospitable wastes of tussock, lichens and bare rock above the bushline, mainly because so few animals are adapted to live there. In other parts of the world the most successful are the birds and small mammals, but New Zealand is sadly lacking in the latter.

Even birds (in a land of birds) are not as well represented here as in alpine regions elsewhere, but what New Zealand lacks in variety, it makes up for in its exciting peculiarities. The native parrots of New Zealand, for instance, are all endemic. The kiwis are unique, and some of our ducks are comparatively rare. The takahe has, several times in the last 100 years, been considered extinct, but about 200 pairs continue to survive in a tiny portion of the Fiordland National Park.

The only true birds of prey native to this country are the two hawks; one a falcon, the other a harrier. The New Zealand falcon is found nowhere else, but belongs to a family of 38 species distributed throughout the world. Although not common, it is most likely to be found in isolated back country valleys. It is smaller, darker and more streamlined than the harrier, striking down smaller birds in flight, and diving when in pursuit at speeds in excess of 160

Left: *Kea and falcon country at the headwaters of the Waimakiriri River.* Above: *The New Zealand falcon is a bird of the back country and steep, bush-covered gullies. It is more commonly known to high country farmers as the sparrow hawk.* Far right: *Found only in the South Island, the habitat of the kea is in the upper limits of alpine forests and the low scrub and rocky faces of the high tops.*

km/h. Its prey consists mainly of small introduced birds and their young, mice, rabbits and even occasionally, a stoat.

Little evidence of their presence is noticed during the breeding season until the young are hatched. But secretive though they are, a likely intruder will be attacked fearlessly in a display of aggression worthy of the larger and more famous birds of prey of other lands.

In contrast, the Australasian harrier is uncommon in the high country and in heavily forested areas but abundant elsewhere. It feeds mainly on carrion from the roadside, insects, lizards and occasionally small birds.

Kea and Kakapo

In New Zealand there are three native parrots (as distinct from parakeets) of which two, the kea and the kakapo are high country birds.

The kea, commonly called the mountain parrot, is restricted to the high country of the Southern Alps, Nelson and Marlborough, where it

takes the place of large birds found in mountain regions of the continents, though it is not primarily a bird of prey.

It is a strong flier and a powerful bird, noisy, exuberant and inquisitive. Often described as mischievous, this highly adaptable parrot is a source of amusement and annoyance to many a tramper and camper. Though accused of molesting sheep for many years, the evidence was largely circumstantial and the bounty on its beak was eventually lifted and the bird given protection in National Parks. More obvious were the antics performed upon unsuspecting victims: unlacing of boots and tents, or the rending of fabrics, or even glissading down iron roofs of alpine huts.

Its colouring is a striking combination of olive-green and scarlet, the latter showing to advantage on the underside of the wings when the bird is landing.

The kea spends much of its foraging time on the ground, even when in the mountain forest. Stones are turned over with its strong, hooked beak, and the larvae of large alpine beetles form a substantial part of its diet. Its nest is either a burrow extending beneath the roots of a large tree, or a cavity amongst a pile of loose rocks.

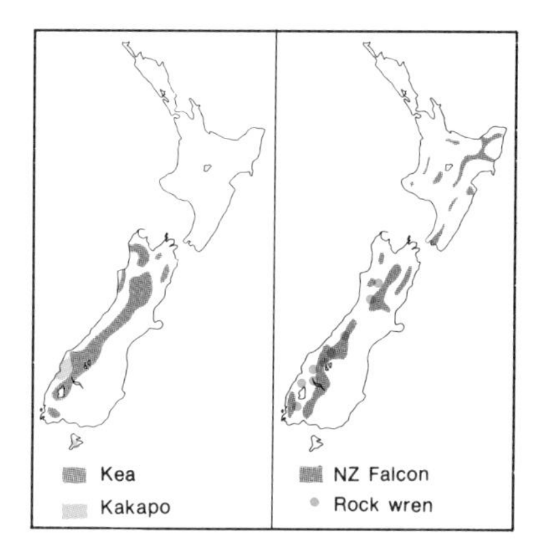

Extremely rare and known only to exist in a remote and limited part of Fiordland, is the largest of our native parrots – the flightless kakapo. It is not particularly colourful by parrot standards, but its attractive shades of green give it a remarkable camouflage in its natural habitat of forest or subalpine scrub. It is described in more detail in another chapter.

The kakapo is a bird of the mossy beech forest and the alpine meadows above the timberline.

The Takahe

The takahe is a large flightless bird standing approximately 50 cm, of somewhat similar appearance, but more massively built than the familiar pukeko, with blue plumage on head, neck, breast and abdomen; the back, rump, tail and wing feathers olive green, the under tail feathers white and the bill and feet red. Both male and female are similar in appearance, although the female is slightly smaller. They are unique to New Zealand and are one of the rarest birds in the world. The takahe is found predominantly in the Murchison Mountains in the South Island, a region of narrow glaciated valleys and steep peaks rising to between 1,500 m and 1,800 m above sea level.

The vegetation is beech forest to an altitude of 1,000 m and then alpine grasslands, dominated by tall snow tussock species. For the greater part of the year the takahe inhabit the grasslands where they feed on tussock, fern stalks and other plants. They are voracious feeders, requiring areas up to 56 ha, of which they are highly territorial and which each pair of birds will actively defend. They lay eggs between mid October and late December in nests situated in well-drained areas above the bush-line and well concealed among tall tussock and scrub which form a bower with entrances to the front and rear of the nest. One to three eggs are

Unique to New Zealand, the takahe is found predominantly in the Murchison Mountains of the South Island.

laid. They are buff coloured with mauve and brown markings. Incubation lasts 30 days and is shared by both male and female. The chick is jet black with short fur-like down, reddish-brown legs and a shiny black, white tipped beak.

By the time the chicks are one month old they have acquired their juvenile plumage which is similar in colour to the adult and have changed their initial diet of insects to one predominantly vegetarian.

Most takahe remain in pairs for life. If one member dies the survivor will acquire a new mate but will not breed until the season after mating.

Several times during the last 100 years the takahe has been considered extinct. After the discovery of subfossil bones in 1847 only four sightings of live birds occurred in the next 50 years, and none at all in the following 50 years. Then in 1948, a small population was discovered in a remote valley on the western shores of Lake Te Anau. Since then about 200 pairs have been found in an area of about 65,000 ha which includes the Murchison Mountains – now designated a restricted area and closed to all except those involved in official research or deer control.

The Rock Wren

Slightly larger than its cousin the rifleman, but smaller than a sparrow, this nimble little bird is one of the few species able to survive among the screes, boulders and alpine scrub high in the Southern Alps. It forages for insects among the vegetation and also eats the fruit of alpine plants, in a characteristic bobbing up and down, knees-bend performance, only taking to the wing when necessary.

Nests are built in crevices, under boulders or in dense shrubs: how the wren survives when snow blankets its habitat is something of a mystery.

Sparingly distributed along the damp, mountain beech timberline of the Southern Alps is the closely related bush wren. A small greenish bird, it is now rare on the mainland and its status on Stewart Island is doubtful. In Fiordland, where it comes down almost to sea level, it may easily be mistaken for the rifleman. Here it is sometimes seen feeding on the ground as well as in the trees, bobbing its head in its characteristic way.

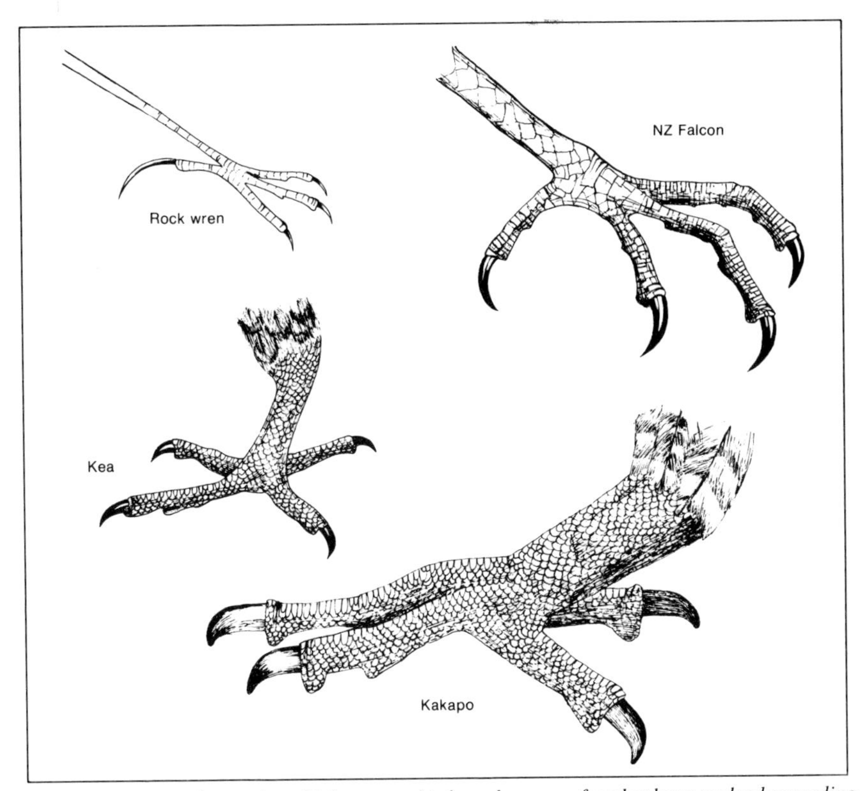

Special feet for special uses: these high country birds each possess feet that have evolved according to their habits and habitat. The New Zealand falcon is a true bird of prey with sharp talons and a strong grip; the rock wren with claws that can cling to the smoothest rock faces; the kea, suitably adapted for hunting on the ground or in the air, and the kakapo, strong and heavy for walking and tree climbing.

A male rock wren at its nest entrance in mountain flax.

Arthur's Pass National Park

The Arthur's Pass National Park is a rugged and mountainous area of about 100,000 ha, situated in the centre of the South Island. It is a land of jagged skylines, tall snowy peaks and snowgrass-clad ridges, deep gorges, steep bushcovered hillsides, sheer bluffs of angular black rock, high waterfalls, wide shingle riverbeds and rushing torrents. The Park is traversed by large rivers; on the east the Waimakariri and Poulter and their tributaries draining to the Canterbury Plains, and the Taramakau and Otira with their tributaries draining to the west.

Most people regard the Park as a region of abundant forest cover, but the impression gained from the valley floors is somewhat misleading, for the area of mountain ridges above the bushline approximately equals the forested area. From valley floors at about 700 m on the east the peaks rise to 1800 m or more, but the western valley floors lie at approximately 300 m. The views from these high peaks range across ridge upon ridge of broken rock, with snowfields lying beneath them and deep-cut valleys below. Within the Park lie ten named peaks over 2100 m in height, and twenty-one over 1800 m.

Many colourful legends have been woven around the uncanny agility of the chamois in its almost inaccessible alpine habitat.

The fauna of the park is obviously divided into native species and those introduced by man. Native birds, insects and reptiles have long been present and a close association and interdependence with their preferred habitats is apparent. Most noticeable are the birds; a great variety of the more common species is evident, and some not so common. Broad shingle beds and valley flats are favoured by pipits, paradise ducks and dotterels. Pipits, running along the ground or fluttering short distances, are friendly little birds; paradise ducks are more wary. A familiar overtone to the dull roar of the river is the squawk of the white-headed female and the honk of the black-headed male as they wheel overhead or, dragging wings, try to distract would-be foes away from their young. The open alpine areas are kea country and forests conceal tuis, bellbirds, pigeons, fantails and tomtits.

Blue ducks are present in the gorges of forested creeks, kiwis are often heard at night but rarely seen, and flocks of yellowheads squabble in the trees accompanied by chattering parakeets and kakas searching for insects in decaying timber.

Introduced animals have modified the natural balance considerably, resulting in much damage to forest and alpine grasslands. Sheep arrived first, followed by goats, red deer, chamois, hares and opossums. However, alpine life is the most diverse among the invertebrates. Cicadas, grasshoppers and moths are numerous. Many kinds of flies including crane-flies, bristle-flies and dragonflies are common. Bugs, butterflies, weevils and beetles, wetas, cockroaches, harvestmen, centipedes and millipedes are also present in large numbers, but often unseen. Swift hunting spiders are common, as is the prominent nest of the nursery-web spider, in grassland and scrub.

Although more common in the lowlands, the New Zealand nurseryweb spider is frequently encountered high up in the mountains where nurseries bind together the tips of tussock grasses.

The desolate magnificence of Arthur's Pass.

And a large orange spider-hunting wasp searches for its prey in the tussock.

Rivers and streams are inhabited by brown trout feeding mainly on the larvae of caddis, stone and mayflies. Under the stones in the more stable streams hide smaller fish, the fast-swimming galaxias and slower bully. The pools of forest bogs are inhabited by an assortment of fascinating invertebrates including giant dragonflies, water-boatmen and back-swimmers, diving beetles and the larvae of the bright red or blue adult damselflies which flit over the water surface.

The banded dotterel is New Zealand's most numerous dotterel. It is a very adaptable bird and will nest almost anywhere, but stony river beds appear to be first choice.

INLAND WATERS

New Zealand is a country with a relatively high rainfall but little more than 30 per cent of the rain that falls on the land finds its way into rivers. The rest either accumulates as ponds and lakes, evaporates or is absorbed by porous soils. Such collections of water, be they lakes or rivers, are very important. They provide fresh water for drinking, watering livestock, irrigation and transportation. They harbour an important food source in fish, support myriad life forms that have evolved for survival in these specialised habitats and could exist nowhere else, and not least, inland waters possess a unique natural beauty.

Lakes provide quite different conditions from those of swift-moving rivers. Oxygen, the most important of the many gases, salts and minerals dissolved in water, is necessary for the respiration of all animal life. Because of its very nature, a moving river generates its own oxygen content through turbulence; and the cooler the water the more oxygen it is capable of absorbing. In a lake, the oxygen enters at the surface, either directly from the atmosphere or as a by-product of plant respiration. In summer, the sun heats the calm surface water, forming an upper layer called the epilimnion which floats on the denser cold water – the hypolimnion. As the surface water gets warmer, an intermediate layer, the thermocline, is formed, through which the temperature decreases rapidly. These three layers do not mix, so the bottom receives no oxygen, though it does receive a constant rain of organic debris from above. Then, in winter, the surface cools and the water circulates until it is thoroughly mixed, with oxygen evenly distributed throughout, and dissolved nutrients from the bottom are returned to the upper layers.

It follows that most lake life is confined to the shallows but fish, especially large fish, range through all levels of the lake. In rivers, the inhabitants are dependent on a food supply which in turn is influenced by the rate of water flow and the nature of the bottom. Slow water usually has a sandy or muddy bottom and is one of the less favourable habitats; fast water contains a large variety of living conditions and plenty of stones and crevices that provide shelter. Torrents, where the water is very oxygenated, are usually rocky and steep and most potential food sources are quickly washed downstream. However, very few insects are able to resist the force of water in this type of habitat. So, at each stage of a river's journey the character of life changes. An animal, whether it be a fish, an insect or a bird, which can tolerate the slow-moving, silt-laden water of the lower reaches would find life impossible in the fast flowing upper reaches.

Fish

About one third of the 26,000 known species of fish live in inland waters, eating an immense variety of food which ranges from nutrients in suspension and sediments in mud to aquatic insects, plants and even other fish. In New Zealand there are fewer than 50 species of freshwater fish, of which 23 are endemic and four native. The remainder are introduced species, some liberated to provide sport fishing, like the trout, and some like the sailfin molly probably introduced as aquarium fish. In comparison with other countries New Zealand has a fairly small number of species and most are poorly known. Britain, for example, has a similar number of which a large proportion are popular sporting fish; Japan with an area only slightly greater than that of New Zealand, has about 130 species.

Trout, Salmon and Char

Many people in New Zealand would regard the Salmonidae family – trout, salmon and char – as being more or less our total representation of freshwater fish, perhaps tentatively adding eels and whitebait to the list. When the European settlers arrived in these parts in the last century they brought with them the eggs and fry of salmon and trout in an effort to re-create the sporting opportunities of their homelands. Success was slow but eventually some species became established.

First the acclimatisation societies, then the Government, took on the responsibility of liberating imported species but too often the fish were released in areas only marginally suitable where some species failed altogether and others met with only limited success. Those that prospered most were the brown and rainbow trout and the quinnat salmon.

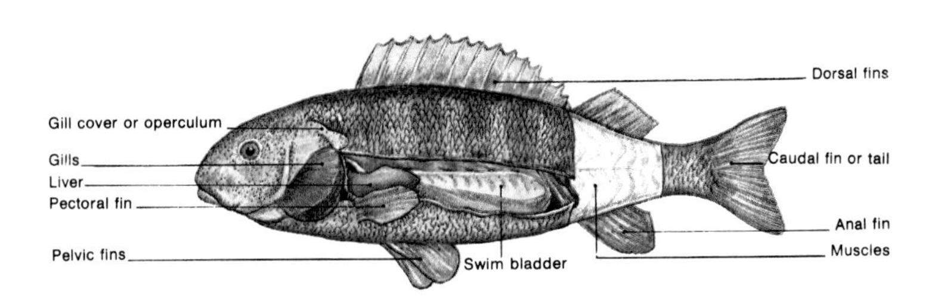

A typical bony fish cut away to show the internal organs. The gills extract oxygen from the water and the gas-filled swim bladder provides the buoyancy which allows the fish to rise and fall in the water. The fins and tail stabilise and propel the fish.

The brown trout is the most widely distributed of the introduced species and in some districts is relatively common. It is found in nearly all river systems of both main islands south of the Coromandel Peninsula. North of the Coromandel the water is generally too warm. Brown trout are fiercely territorial and a large fish will often take over a pool for itself chasing others away. The availability of space and cover determines the number and growth of the trout.

The rainbow trout is distinguished from the brown trout by the shade of pink along its sides, the back is a paler green and the tail is spotted. Rainbows are mainly lake fish; however, both species spawn in the rivers and where they cohabit, the rainbow usually dominates. In North America where the rainbow is a native, it is a sea-run fish, but in New Zealand, when they swim out to sea, they tend not to return. When first liberated in this country the rainbow grew to a very large size but have now steadied to an average 50-60 cm and a weight of 2.5 kg. Several North Island lakes – Rotorua, Taupo, Tarawera and Waikaremoana – have good populations: the fish spawn in the tributaries and make their way back to the lakes. In the South Island the rainbow is closely associated with high country waters such as Lake Te Anau and Lake Manapouri and the Waimakariri and Rakaia Rivers. Hatcheries in both islands continue to restock the lakes.

Trout and salmon belong to the same family and their life histories are similar. Some species spend part of their lives in salt water and others do not, but they all need freshwater to spawn. The eggs are quite large, about 8 mm, and are discharged along with the milt of the male in a depression in the gravelly bed of a swift running stream. The depression, made entirely by the female, is called a redd. Incubation takes several months. In some species the young immediately head downstream and out to sea; in others they take several years to cover the same journey, and some stay in the river of their birth.

Rainbow trout (above) *forms the basis of lake fisheries, especially in the North Island where considerable success has been achieved around Rotorua. They also occur in most upland lakes in the South Island. The brown trout* (right) *is amongst the best known fishes in our waters, widespread throughout and common in many areas. Brown trout forms the basis of the important river and stream fisheries of the southern North Island and entire South Island.*

Of the three salmon species present in New Zealand the quinnat or Pacific salmon is the most common sporting fish. The young quinnat may take anything from 3 to 18 months to reach the sea where it stays for one to four years before commencing the long struggle back. In New Zealand waters the adult quinnat grows to 70-80 cm and weighs 6-8 kg, prior to the migration, which starts in December. By March the fish are entering the river estuaries and are still in good physical condition. But feeding now ceases and by the time they reach the spawning areas in late April their condition has deteriorated remarkably. After spawning, their goal achieved, both male and female will die.

The main quinnat salmon rivers are those of the South Island's east coast between the Waiau in the north and the Clutha in the south. Quinnat have also become established in a few rivers in south Westland but stocks are not yet sufficient for fishing. Self-sustaining, land-locked populations are smaller than the sea-run fish but have the same life history, migrating from river to lake instead of to the sea.

Sockeye salmon occur only in the Upper Waitaki River system, Lakes Ohau, Benmore and Waitaki. They were introduced in 1901 in the hope that a sea-run population would develop. As this did not eventuate, it seems probable that the original ova were from land-locked populations. It is a small fish in New Zealand, not utilised as a commercial or sport fish and little is known about its habits. In its restricted habitat it is preyed upon by trout and char.

The Atlantic salmon is another land-locked species. Though fairly well established in the Te Anau-Manapouri system, the Atlantic salmon has been a disappointing introduction. A great fighter in its natural European habitat and growing to 150 cm and 35 kg, the Atlantic in New Zealand has only attained an average 60 cm and 2 kg. It is a fish that will leap nearly 2 metres up a waterfall and swims strongly against the swiftest current. Unlike the quinnat salmon, the Atlantic may live on to spawn as many as four times. Since the 1920s management of the Atlantic salmon has been largely neglected and rainbow trout, recently liberated in to the same system, are taking over from the salmon.

Quinnat salmon (top) *occur mainly along the east coast of the South Island and all the major rivers east of the Main Divide. The sockeye* (above) *occurs only in the lake-limited form and is much smaller than the other salmon. The larger one in the picture is the brightly coloured spawning male, the smaller is more typical.*

Brook char have a preference for small inaccessible streams which makes it almost certain that their distribution is wider than records indicate.

Two species of char – related to the salmon – have established themselves since their liberation. The brook char, released in 1877, is an attractive dark olive-green fish with yellow spots on its sides and reaching 20 cm in length. Although introduced as a game fish it migrated upstream to small almost inaccessible waters where it is not often seen. The mackinaw was liberated in 1906 in Lakes Pearson and Grassmere. Lake Pearson is warmer and shallower than its normal habitat but the mackinaw survived and now maintains a self-sustaining population. Their life span is long, often exceeding 15 years and although not normally growing to a large size, a specimen of 3.2 kg has been recorded.

Rudd

Rudd is a more recent, and illegal, introduction, and has been reported in several localities north of the Waikato River. They are carnivores, feeding mainly on insects and crustacea on the surface. Large rudd have been known to include smaller fish in their diet. Like perch, the rudd lacks predators and unless they are managed as a resource, the rudd populations will also consist of under-size fish.

Carp

Most of our introduced fish were originally liberated in the hope that sporting opportunities would follow, but some have arrived for other reasons. During the last decade Chinese grass carp have been released into the lakes of Rotorua in an attempt to control the water weed. Unfortunately, they seem to be responsible for less desirable effects including damage to the invertebrate fauna. Other fish have appeared, probably released by aquarists, as stock, though no actual records are known. The common goldfish is one of these. They like still water and are able to survive in stagnant pools with very little oxygen. Unlike the specially bred varieties, these 'wild' goldfish are olive-bronze to deep gold without the flamboyant fins; only occasionally do the familiar red-gold or pearly specimens appear.

Tench and Perch

Although the trout and the salmon are our famed sporting fish several other species have also been liberated in our lakes and rivers for game. Tench, introduced in 1867, is a seldom seen species that prefers warm, shallow lakes or gently-flowing, weedy streams. It is found in Northland, near Otaki, near Oamaru and the West Coast. Perch arrived about the same time as tench, enjoys similar waters, but is more widespread. Both these species are regarded as good sporting fish in the Northern Hemisphere but are largely ignored in New Zealand. Because of the large number of eggs laid and the absence of predators, the perch populations tend to be proliferations of small fish – too many for the available food.

Live Bearers

The sailfin molly is a warm water species introduced to some thermal lakes of the North Island's central plateau, and will not survive low temperatures. It grows to about 6-7 cm and the male develops a very prominent dorsal fin. This attractive aquarium fish, and the related mosquito fish are the sole representatives in New Zealand of the Poeciliidae family – the live bearers. The eggs are fertilised within the female and the young are born live at almost any time of the year. Mosquito fish were introduced in an attempt to control mosquitoes by biological means. Of several initial liberations only those in Northland still remain.

Galaxiids

The largest family of native fish are the galaxiids which are ubiquitous in running water. The family contains 13 of the 27 native freshwater species found in New Zealand. Most galaxiids are small fish between 4 and 15 cm long, although some species reach 25 cm and one, the giant kokopu averages over 30 cm. All have similar features such as smooth, thick leathery skin with no scales, frequently covered with a coat of mucus. Many of these fish are very sensitive to change within their habitats and even the slightest changes cause dramatic fluctuations in their populations. The banded kokopu, short-jawed kokopu and the koaro are species that require good cover such as that provided by overhanging banks and low vegetation of native forests. Where forest cover has been removed these species are no longer found. Apart from the general need for cover the habitats of these species vary considerably. For example, the banded kokopu prefers small, stable streams with rocky beds and small pools and is sometimes found in forest swamps; the short-jawed kokopu also favours small pools in dense, unmodified native forest; the koaro likes a tumbling stream and is rarely found beyond the bush. Each of these species is important as part of New Zealand's wildlife heritage but also because the juveniles, along with those of the inanga and the giant kokopu, form the basis of our whitebait fishery.

The adults of the forest dwelling species spawn where they live and the larvae move downstream and out to sea in the autumn. The adult inanga, itself however, goes down to the estuaries where it lays its eggs among the reeds and grasses of the riverbank. This event always coincides with a spring tide; when the tide falls the eggs are washed down to the base of the grasses where the constant dampness prevents them from drying-out. Two weeks later, at the next spring tide, they hatch and the fry are washed out to sea as the tide recedes. They spend the whole of the winter at sea before beginning the springtime migration back up the rivers.

Whitebait of some species are very adept at climbing and will wriggle their way up steep and swift-flowing rivers, waterfalls and vertical barriers, having already run the gauntlet of larger fish, seabirds and man's nets in the estuaries.

Dwarf inanga are found only in a few lakes in North Auckland whilst alpine and long-jawed galaxias both

Top: *The perch is an attractive fish with a widespread distribution in lakes and gently-flowing, weedy rivers. Survival for the sailfin molly* (above) *depends entirely on the thermal swamps at the southern end of Lake Taupo.*

Two galaxiids of very different habitats: top: *the giant kokopu, a large secretive fish of swamps and overgrown pool margins, and,* above: *the small inanga, the best known contributor to the whitebait catch.*

prefer the cold, fast-flowing, gravelly streams of the Southern Alps. The common river galaxias grows to about 15 cm and is confined to the cool tumbling streams of the upland areas of Canterbury and Otago.

Mudfish

The mudfishes are an interesting group of galaxiids that have adapted to a specialised existence in swamps and drains that tend to dry up during summer. There are three related species in New Zealand: the Canterbury mudfish, the brown mudfish or spring eel, and the black mudfish, living respectively in Canterbury, the southern North Island and Westland, and the swamplands north of Auckland. Mudfish survive the dry months by aestivating (maintaining a state of torpor) in the mud or forest litter. They can burrow to a considerable depth, the brown mudfish being found at 2 m below the surface. Few populations are known of the Canterbury mudfish and its future looks bleak. Its natural habitat can only be supposition as the Canterbury Plains have become so modified and little is left of the original forest. A few small, swampy creeks and drains are the remaining sources of this rare and interesting species.

The black and brown mudfishes are similary elongate, tubular fish but lack the pelvic fins of the Canterbury mudfish.

Eels

Two species of eels are resident in New Zealand: the short-finned eel which is also found in Australia, and the long-finned eel which is endemic. Both are to be found in most river systems; those far upstream usually being the females and those that stay close to the estuaries are usually the males. The females of both species are considerably longer than the males. A long-finned eel can attain a length of nearly 2 m, although average is about 120 cm and 65 cm for female and male respectively.

Eels take a long time to mature: up to 47 years for the long-finned female and up to 34 years for the short-finned female. During the year of their maturity they change physically, the head becoming slender and tapered, the fins and eyes enlarged. Then in late summer they cease feeding and leave the rivers for the open sea.

Where they spawn is still a mystery. It is surmised that they congregate in very deep water somewhere in the Coral Sea, dying immediately after laying their eggs. On reaching the surface the eggs develop into transparent, gelatinous, willowy larvae called leptocephali, which drift with the currents and the tides until they reach New Zealand. A remarkable transformation then takes place. The gelatinous body is absorbed, the length reduced and the young eel becomes a slender, transparent glass eel about 6-7 cm long. Thousands then commence their journey into freshwater and swim upstream in company with even greater numbers of whitebait.

Colouration soon appears – grey brown on back and sides, pale brown underneath, as they develop into elvers. No obstacles impede their progress, steep waterfalls are climbed, huge dams overcome. They will even cross land, when the grass is damp, to reach waterholes and drains that have no connection with running water. They hide under rocks and gravel in the stream bed when small, moving to the shelter of overhanging banks as they grow

Resembling some species of Galaxias, *the Canterbury mudfish has adapted to a specialised life in semi-permanent waters – bogs, creeks and drains that tend to dry up in the summer.*

A short-finned eel slithers over damp grass on its way back to the sea.

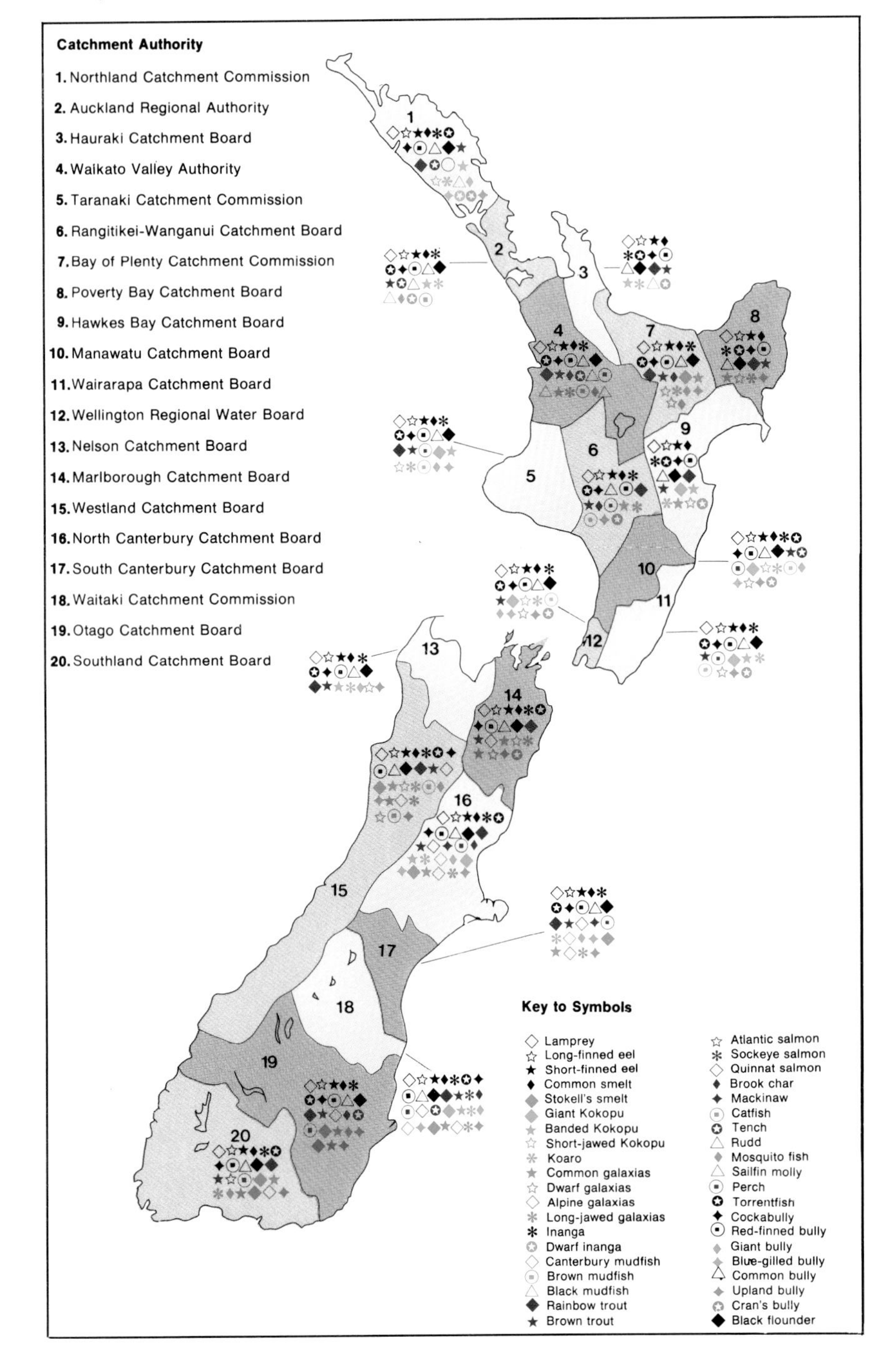

larger. The population in any single pool depends on the amount of cover available.

Smelt

Smelt are found throughout New Zealand. The more common of the two species is a deep-bodied, slender fish, bright silver with a purplish sheen and large eyes. Very similar is Stokell's smelt which is restricted to the rivers of Canterbury, Otago and Southland coasts. Both migrate upstream to spawn, preferring quiet, gently-flowing pools with a sandy bottom. Then the larvae are washed back to sea where they spend most of their lives. Lake populations do not go to sea but provide trout with an important source of food.

Bottom Feeders

Little known among our freshwater natives is the torrent fish which, as its name suggests, favours rapidly-moving water. It is one of the few fish that seems not to have been affected by the clearing of the bush. Its mouth is positioned well under the head which suggests that it feeds on the bottom of its gravelly and cascading habitat. Six species of bully are also bottom feeders; all of them stocky with blunt rounded heads and tubular-shaped bodies. The giant bully will often measure over 10 cm, occasionally over 20 cm, while the smallest, the blue-gilled bully rarely exceeds 7 cm. The latter is often found in the same habitat as the torrent fish. Most spectacular of the bullies is the male of the red-finned species, whose lateral colouring of bright orange-red bands and blotches is superimposed on a dark grey-brown background. The fins are large and fleshy, flecked with red and the dorsal fin is outlined with an iridescent blue-green. The underneath is paler with the lower side of the head a vivid

Possibly the most attractive of the native freshwater fish, the male red-finned bully becomes a deep velvety black when spawning in late winter and spring.

Large barriers like the Raukawa Falls on the Wanganui River are often no obstacle to eels and spawning fish intent on swimming upstream.

lemon-yellow. These colours are intensified in deeper water, but the female remains a comparatively dull olive-green.

Four of the bullies spend part of their lives at sea; the other two, Cran's bully and the upland bully are wholly freshwater species. Most adaptable of them all is the common bully, a fish that is quite at home in lakes no matter what kind of bottom, or by river margins with overhanging banks and fallen logs. In some areas it has been introduced into lakes as potential trout food.

Crustacea and molluscs

Apart from fish, our rivers, lakes and wetlands contain a wealth of animal life including amphibians, insects, molluscs and crustaceans that exists solely because of the presence of water.

Crustaceans and insects both belong to the invertebrate group of animals called Arthropoda and form the largest phylum in the animal kingdom. Their bodies are segmented and encased in a horny layer, or cuticle, which may be flexible or rigid and forms an external skeleton. Crustaceans differ in that they are primarily aquatic, breathe through gills and have two pairs of antennae.

Most freshwater crustaceans in New Zealand are small and insignificant yet seven of the eight subclasses are represented. The greatest importance of these tiny organisms is their role in the food chain, converting detritus, fungi and bacteria into food for other animals. Most of the lesser crustaceans are omnivorous – feeding on anything available.

Crayfish and shrimps are probably the best known of our freshwater crustacea. Both are found throughout the lowland streams and rivers of New Zealand as well as in lakes and swamps. The crayfish prefer a bushland stream with a soft bed to burrow in although they will sometimes inhabit a rocky stream using large rocks as shelter. But they are not confined to areas of permanent water. In some parts of Westland they are known to live in bogs, burrowing into the mud in summer when surface water dries up.

Freshwater crayfish resemble closely the marine variety except for a pair of limbs with powerful pincers located behind the antennae. In this respect they are more akin to the European lobsters. Like all crustaceans, the crayfish sheds its shell as it grows, the adult moulting once a year. Shortly before moulting the shell becomes soft as the calcium salts are absorbed into the body and stored. The moult always takes place under cover as the crayfish is unprotected and very vulnerable at this time. It takes about ten days for the new shell to harden while the secreted lime is reabsorbed from the body; but the pincers, needed for defence and catching food, harden in only five days. Crayfish are preyed upon by many fish, particularly eels and trout. Both the freshwater crayfish and the transparent shrimp are active mainly at night, which is the best time to search for them.

Molluscs are unsegmented animals, with body cavities and highly developed blood and nervous systems. Their body is divided into a head, a muscular foot and a humped back covered by a mantle of skin which is folded to form the cavity used as a lung in some forms. Everyone is familiar with the mollusc remnants that litter the seashore: bivalve shells, limpets, chitons, and so on, and with the common terrestrial forms of slugs and snail, but those that depend on freshwater are not so well known.

Thirty-one species of freshwater mollusc are to be found in New Zealand. Mussels are the largest.

The Lamprey

The most primitive of the fish-like creatures of New Zealand's inland waters is the lamprey. Following European settlement the lamprey has become less common and because it is nocturnal, is now rarely seen. Although rather eel-like in appearance the lamprey is not a true fish, and differs in several ways. It has no bones but a skeleton made of soft cartilage, the backbone being a flexible rod. It has no jaws, but a sucking disc armed with rows of sharp teeth; its tongue is piston-like, the tip of which is also armed; and it possesses keen vision. These are characteristics that come with maturity, like the deep blue stripes on a startling silver background, but for most of its life the lamprey is a dull, drab grey-brown.

Young lamprey, called ammocoetes, are found in muddy backwaters and sandy shallows and can move very swiftly, burying themselves in the mud if necessary. When maturity is reached the new adult makes its way to the sea where it will feed by clamping its sucker-mouth on to the side of its prey, (usually larger marine fish) rasping a hole in the flesh and living off its tissues and body juices. It is not known how long the lamprey stays at sea, but eventually it returns to the rivers to breed.

They are usually dark-brown or black on the outside and pearly within, and grow to a length of 12 cm. The largest freshwater limpet is a luminous animal with a black shell. It grows to 10 mm and may be found firmly attached to stones in swift water. Two other limpet species, one confined to the North Island, the other to the South, grow to a mere 4 mm and prefer sluggish water.

Snails are numerous in our inland waters, but are mostly very small. One of our largest species is the decapitated snail which grows a 3 cm brown-banded spiral shell. It is widely distributed, particularly near the coast, and frequently in brackish water. Also fairly common is the tall, spired dark water snail. A second group of snails with thin, often flat shells are known collectively as the pond snails. Practically all the freshwater snails in New Zealand are exclusive to the area and not found anywhere else in the world.

The flat, worm-like animals, also unsegmented but lacking a true body cavity are the flatworms or planarians. They are usually less than 25 mm long and instead of swimming in the water they glide across solid objects or underside of the surface film. Planarians are carnivorous, either catching their prey with the aid of a slimy secretion, or feeding on the dead bodies of larger creatures. Some planarians have the remarkable ability of regenerating lost parts of the body, and they multiply simply by breaking into two parts.

Insects

Many insects use water as a breeding medium for eggs and

Two species of freshwater crayfish are found in streams and lakes throughout New Zealand. This one is Paranephrops planifrons *of the North Island, Marlborough, Nelson and Westland districts.*

The largest and most distinctive of New Zealand's dragonflies, sometimes called the devil's darning needle.

Several water-skaters of the genus Microvelia *feeding off a floating dead moth.*

developing larvae; some live all their lives underwater whilst many adults become aerial. Dragonflies, damselflies, mayflies and stoneflies all spend their early life in freshwater, the first two emerging as strong-flying adults, the latter two much weaker in flight. Eight dragonflies are native to New Zealand. They are large, fast-flying, stout-bodied, and strongly territorial. The most easily recognised dragonfly is the black and yellow devil's darning needle with a wingspan of up to 13 cm. Damselflies are shorter, slender by comparison and brightly coloured red and blue. The eggs of both are laid on the banks of streams or below water on swamp vegetation where there is some water movement. The nymphs construct tunnels to live in and catch unwary insects that pass by. As long as four years may be spent underwater before progress towards the adult stage is obvious.

Mayfly and stonefly nymphs are frequently found together but can be distinguished by the number of tail filaments – stonefly two; mayfly three. Both are common under stones in unpolluted streams. Vast quantities of eggs are laid by the mayfly but relatively few reach maturity. Trout is the major predator although other aquatic insects, including the stonefly nymph, consume a great number of the eggs. Adult mayflies live for only a few hours, never more than a couple of days.

Other insects whose larval stages are dependent on water include cascade flies, harlequin flies or midges, shore flies, caddis flies, sandflies and mosquitoes. Of the 12 species of mosquito in New Zealand, none fortunately, is the malaria carrier of other countries, although one West Coast species does transmit a flu-like virus, originally amongst birds but now to humans also. Females of all species, however, feed on animal blood which is necessary for the maturation of her eggs. The male mosquito does not feed at all.

Insects which spend their whole lives underwater are as varied as their aerial cousins. Back swimmers, water boatmen and diving beetles all need oxygen to respire, replenishing their supply by breaking the water surface and 'capturing' air which they store in many parts of the body. Beetles carry their air supply by trapping a bubble beneath the wing-cases, which allows them to remain submerged for several minutes.

Surface-water insects include pond skaters, shore bugs and water measurers, which leave behind them a V-shaped wake as they walk across the water. The dark scum-like patches sometimes found on the surface of ponds and slow-moving streams are probably colonies of springtails, which have congregated there from nearby leaf mould.

Introduced Frogs

Most amphibians are animals of both land and freshwater and the most common frog in New Zealand is an Australian import which depends entirely on a watery habitat. The green tree frog, introduced in 1867, is found in streams, ponds and marshes throughout the country. It grows to about 8 cm and although some golden specimens have been found it

The green tree frog frequently leaves its watery habitat at night, loudly announcing its presence from September onwards through the summer months.

is usually green. There is some variation within the species ranging from smooth skin with irregular gold spots to those with a warty skin, olive in colour, and a green dorsal stripe. The tadpoles are vegetarian, while the adults feed on a variety of insects. They in turn, become part of the food cycle by providing many water-birds with an important part of their diet.

The other introduced frog is an inconspicuous little fellow that is common in Westland, the Waiau River system of Southland and other more remote parts of the South Island. Called the brown tree frog or whistling frog, this species was introduced from Australia in 1875. It is only about half the size of the green tree frog and the colour varies from a uniform dark brown to a dusky grey or fawn, either with small dark flecks or no markings at all. Like the green tree frog it is mainly nocturnal but occasionally makes an appearance on wet or overcast days.

Three native frogs which are not dependent on the presence of free water, are discussed elsewhere.

Birds of the Wetlands

Swamps marshes, rivers and lakes, in fact, all wetland areas support a great variety of bird-life such as ducks, rails, bittern, herons, fernbird (in drier parts with bracken) as well as introduced swans and geese.

All ducks, swans and geese belong to the family Anatidae; all have short legs, webbed front toes and broad bills and are usually gregarious. The shoveler, grey duck and mallard are known as dabbling ducks from the way they feed. They are also the three major game ducks in New Zealand. The New Zealand shoveler, though not abundant, is widely distributed wherever extensive lowland swamps are a feature of the countryside. It is a very fast flier and generally most mobile just before dawn or after dusk. In common with many other birds only the male shoveler is colourful, the female being quite drab in comparison. This drabness is a great advantage in that it offers near-perfect camouflage to the nesting female and is some protection against hunters.

The male and female mallard are similarly contrasting, but the sexes of the grey duck are indistinguishable. Grey duck is found

The male shoveler (right) *is the most colourful of the native ducks whilst both sexes of the grey duck* (above) *are similarly drab.*

throughout New Zealand on rivers, lakes, ponds, swamps and inter-tidal mudflats. Unfortunately it is nowhere near as plentiful as it was 100 years ago because few of the extensive secluded swamps – its favourite breeding habitat – remain.

The nest of the grey duck is often located away from water, well hidden, on the ground or occasionally in hollow trees. The eggs are laid in August, one per day until the normal clutch size of 10 is reached. Incubation takes four weeks and the eggs all hatch within a few hours of one another. The ducklings swim almost immediately and flight is possible in about ten weeks. Once the brood has flown, the adult females seek seclusion for their main annual moult. Both males and females drop their main flight feathers during the incubation period and grow new ones over three weeks or so during which time they are flightless and vulnerable.

Three smaller species, none of them common, are the grey teal, scaup and brown duck. The teal, though sometimes mistaken for the grey duck or the female shoveler and mallard, is noticeably smaller. Its wing beat is rapid and a distinctive white bar on the upper surface of the wings is conspicuous in flight. This self-introduced bird from Australia is very mobile, continually on the lookout for new lakes or lagoons even though there may be sufficient water in the areas they vacate.

Probably the most common of the small ducks and certainly the most handsome is the New Zealand scaup. The male is black with a purplish-green sheen on the head and a bright yellow eye. The female is browner, with a brown eye. In breeding plumage scaup have a small white band on the forehead just above the bill. They are diving ducks and are able to remain submerged for almost half a minute in the large clean lakes that they favour. Their food consists of aquatic insects, molluscs, crustaceans and freshwater plants.

Brown duck (also called brown teal) are now found only near a few swampy streams and ponds in Northland, Fiordland and Great Barrier Island. Designated a rare bird, its habits are little known, except that it is active at night feeding along the shore among the variety of aquatic plants which harbour invertebrate food. Daytime is spent hiding in the thick undergrowth of the creek banks. It dives well and can fly strongly, but rarely takes to the wing.

Though not rare the blue duck is seldom seen. It inhabits the bush streams of mountainous country where the cover is dense and the water swift. A very attractive bird, the entire body, apart from the brown spots on the breast, is a dappled blue-grey. It is a very secretive bird spending most of the daylight concealed in stream-bank vegetation and coming out for a few hours at dawn and dusk to feed. Their crepuscular feeding habits are probably related to the behaviour of their invertebrate prey. Most ducks consume some vegetable material but the blue duck feeds almost exclusively on larvae and adults of aquatic insects such as caddis fly, stonefly and mayfly. These insects emerge at dusk from their watery

hiding places to rest on stones and rocks, thus becoming readily available to the ducks.

In complete contrast to the blue duck, the paradise shelduck is a bird of open pastures and farmland in close proximity to water. Semi-aquatic plants growing in swampy gullies or water-logged pastures are its main source of food, supplemented by the odd earthworm or insect. It is a large duck and in this case it is the female which is the more colourful with its conspicuous white head and bright chestnut body. The drake has dark, variegated plumage. Though an indigenous species it is a successfully managed game bird and many thousands are hunted during open seasons each year.

This Westland bush stream (right) *is a typical habitat of many of the galaxiids, freshwater crustacea and aquatic insects, and of the New Zealand scaup, the diving duck,* (below) *whose chicks are able to dive to considerable depth shortly after hatching.*

Two of the 19 species of grebe occur in New Zealand as breeding birds: the crested grebe and the dabchick. They are characterised by lobed feet (not webbed as the ducks') on legs set far back on the body, and sharply pointed bills. The crested grebe is about the same size as the grey duck but differs markedly by its longer, slender neck, the pointed bill and the crest and ruff on the head. Both sexes look alike and the colourful head adornment is put to spectacular use during the courtship display. It is an expert diver but a reluctant flier. When danger threatens, it invariably dives and covers long distances whilst submerged.

Once widespread, the crested grebe has declined in recent years and North Island breeding lakes seem to have been completely abandoned. Today the species is confined to Westland and east of the Divide as far south as Fiordland. Stragglers are seen occasionally in Lake Ellesmere, Lake Waihola and Southland.

The dabchick is endemic to New Zealand but is closely related to a similar bird in Australia – the hoary grebe. Once common throughout the country, the dabchick is now a breeding bird of North Island lakes

Though a strong flier the blue duck will, when disturbed, make good its escape by diving and swimming out of sight.

it sometimes can be seen feeding on mudflats at low tide, but it is quick to run for cover at the slightest hint of danger.

The marsh crake, a little over half the size of the banded rail but even more colourful, has a yellow-green bill, eyes which are bright red surrounded by a pale steely-blue that also covers the cheeks and breast, and pink legs. The upper parts are reddish-brown with streaks of black and white continuing down the flanks and belly. This retiring bird inhabits both fresh and saltwater swamps throughout the country but is most common north of Auckland and west of the Southern Alps.

Probably the most common swamp bird in New Zealand, though glimpsed only occasionally, is the furtive spotless crake. This small rail is a handsome bird though lacking

and ponds. Like the crested grebe the dabchick seldom flies, but pairs may be seen early in the breeding season skimming over the water surface of new hydro lakes. It feeds by diving in search of insects and molluscs, very small fishes and aquatic weeds, but avoids waters heavily infested with eels.

Swamps and marshlands that fringe many of our lakes, ponds and rivers are the domain of the small rails, the common pukeko and the bittern. The deep blue, red-beaked pukeko is a familiar bird to most New Zealanders and can regularly be seen from the roads in most regions. It is a member of the family of Rallidae, which includes the rare takahe, the wekas and the smaller species: banded rail, marsh crake and spotless crake. Despite the lack of webbing between their toes, the pukeko is a good swimmer, though much of its time is spent fossicking in wet pasture and semi-open marsh for a wide variety of plant matter, snails, worms and insects. It is ungainly in flight but can cover long distances; one bird has been recorded as flying 142 km in three weeks.

The small rails are neither as bold nor as conspicuous as their larger cousins. The banded rail, a miniature version of the weka but more richly coloured, is limited to coastal areas of north-west Nelson and the northern half of the North Island and some offshore islands. The mangrove swamps of North Auckland are its favourite haunt and

The courtship of the crested grebe is most likely to be seen between August and February.

Both members of the family of rails, the pukeko (above) *and the spotless crake* (right) *differ considerably in size and behaviour. The large pukeko is a familiar bird of roadside pastures especially in the vicinity of water. The spotless crake is a small (20 cm) bird and though common in all types of swamp, its secretive, skulking nature allows few sightings.*

the colourful plumage of its cousins. Both sexes are a uniform blue-grey with a deep brown saddle and a striking red eye. Except on offshore islands where ground cover is sparse, the spotless crake prowls the raupo and reeds only revealing its presence by an unusual repertoire of calls. Where even mere vestiges of swamp exist, the spotless crake is likely to be found. Several nests are built of interwoven grass and reeds, always untidy and often over water. One nest is used for laying the clutch of 2-5 eggs, the others probably serve as brooding nests and quickly disintegrate when not used.

Herons

The herons, though closely related to the stealthy bittern, differ considerably in their choice of habitat. Little effort is made to conceal themselves whether present in large numbers or small. Three species of heron breed regularly in New Zealand: the reef heron of the intertidal zone, the white-faced heron, and the white heron or kotuku of Maori legend.

The white-faced heron is by far the most common and the most adaptable of the three. All kinds of wetlands from tidal estuaries to inland swamps, ponds to rivers, are suitable habitats. Its white mask apart, this species is bluish-grey with the faintest tinge of purple; the bill is black, the feet yellow.

Largest of the three indigenous herons is the white heron, or kotuku. Only one breeding area is known in New Zealand, that on the banks of the Waitangiroto Stream near Okarito in Westland, where natural buffer zones provide a refuge for this revered bird. The plumage is pure white in both sexes, legs are black, and eye and bill yellow, except in the breeding season when the bill also becomes black. At this time there are fine dorsal plumes for which the species was once exploited.

As a solitary hunter in ponds and drains, creeks and estuaries, the kotuku seeks a wide range of food: small fish including young eels, frogs, tadpoles and large insects. If food is plentiful enough a single bird may remain in one district all through the winter. Nest platforms, made of sticks, often overhang the water. A clutch consists of three to four pale blue eggs tinged with green.

After the nesting season the white herons of south Westland will disperse all over the country, visiting lakes, coastal lagoons, harbour flats and mangrove creeks.

The Bittern

This sedentary inhabitant of the swamps and marshes belongs to the heron family and is another of the birds that are widely distributed, but not very numerous and seldom seen. Reed beds are their habitat and provide cover from which they rarely emerge. When they do have to cross an open space they make a crouching run and assume once more a motionless pose. In such postures they are very difficult to see, their brown and tawny-buff vertically streaked plumage resembling the reeds, and the bill pointed skywards completing the camouflage.

The bittern is a bulky bird feeding mainly on frogs, eels and other small fish, and insects. In the late evening its characteristic booming sound may be heard, the bird pointing the bill downward and fluffing out the feathers of its neck and chest, thus increasing the resonance of this peculiar bellow.

The kotuku bears a prominent place in Maori folklore, having become a symbol of great beauty, rarity or grace, and also as an inhabitant of the spirit world.

Since 1950 another species of heron-like bird has nested regularly with the kotuku, and apparently nowhere else. The royal spoonbill, self-introduced of Australian origin, is about the same size as the white heron. It is also pure white with black face, and legs but is easily distinguished by its large spatulate bill. Their feeding habits also differ in that the flat bill is used like a scythe on muddy surfaces rather than as a spear in shallow water.

Geese

Long before the European settlement of New Zealand there lived and died three species of native geese; their demise was probably due to their vulnerability, as two of the three were known to be flightless. Of the many species introduced since by acclimatisation societies, only the Canada goose from North America has flourished. Its natural habitat was the prairie grasslands and so it is well suited to the eastern plains of the South Island. The geese increased and spread rapidly, eventually colonising the headwaters of all the large eastward-flowing rivers and their associated lakes from Marlborough to Central Otago. They breed in isolated valleys where tussock grasses are the dominant vegetation, but there is no better place to see them in large numbers than Lake Ellesmere in summer. Geese from throughout the South Island congregate here in thousands to spend together the flightless period of their annual moult. Breeding birds with their young arrive later to swell the flocks before departing the following spring to their high country breeding grounds.

Swans

A swan, also a native of these islands, suffered a similar fate as the pre-European geese and became extinct sometime before 1800. Nowadays, the black swan from over the Tasman is our most common swan, introduced in 1864 to control watercress on the River Avon in Christchurch. More liberations followed and within a

A pair of Canada geese on Lake Wanaka.

few years the birds were observed throughout the South Island, and were ready to occupy suitable habitats in the North Island. Major and permanent colonies flourish on Lake Ellesmere, Lake Whangape in the Waikato and Lake Waihola near Dunedin, while smaller flocks occupy the large harbours of the north, parks in urban areas and small lagoons around the coast.

Swans are vegetarians and feed almost exclusively on aquatic plants, but with more and more wetlands being drained for pasture, natural food supplies have dwindled in many areas. The Wahine storm of 1968 had disastrous consequences for swans at Ellesmere as it devastated beds of aquatic *Ruppia* plants. Populations declined dramatically since and recovery is slow.

Lake Ellesmere also supports the largest breeding population of the white or mute swan. Unstable water regimes and competition for territories on this large area so densely populated with other waterfowl, has prevented this majestic bird from becoming properly established. A small, self-sustaining population exists in Hawkes Bay. The nest of the mute swan is little more than a massive collection of sticks and raupo leaves, usually a long way from its nearest neighbour, but at Lake Ellesmere they are spaced only a few metres apart. Dense raupo affords enough privacy for this unusually gregarious behaviour as long as neighbouring pairs cannot see or interact with one another.

Cygnets of the black swan become progressively darker after successive moults.

Fernbird

Swampy ground, with raupo or bracken fern is also the preferred habitat of the fernbird. Once abundant, this is another of the birds that has suffered with the draining of the wetlands and the burning of fern and scrub. However, where suitable conditions exist this secretive bird is still plentiful. Although difficult to spot in the dense cover of its habitat the fernbird will often betray its presence by its sharp metallic call. It moves quickly trailing its tail through the low vegetation and flies weakly as if restrained by the same untidy tail.

Shoreline Birds

The stony river-beds east of the Southern Alps are frequently used as the breeding ground of several migrating waders, dotterels and gulls. Wrybills, pied oyster-catchers, pied and black stilts, banded dotterels and black-billed gulls are all species that nest in the shingle close to water, and winter in coastal areas many kilometres away. Spur-winged plovers favour similar country especially if associated with wet pasture, and the ever present kingfisher sits patiently in the nearest tree waiting for a fish, lizard or cicada meal.

The Kingfisher

The local race of a species that occurs in Australia and islands in the South-west Pacific, the brightly coloured kingfishers are seen everywhere in New Zealand. Their habitat is open country or forest edge and they are common near lakes, rivers, estuaries and the sea coast. Like the kookaburras, to which they are closely related, kingfishers are less dependent on water than their name implies, and they probably catch more of their food on dry land than they do by diving. Diet varies with locality; kingfishers living near the coast eat a wide variety of crabs, small fish and shrimps from mudflats and rock pools. Away from the sea, worms, insects, lizards and freshwater fish form the basis of their needs. When the opportunity arises they will also sample fruits, take small birds and swoop on an unsuspecting mouse. Nests are made in dead trees or clay banks in a hole excavated by both birds. The clutch of 4 or 5 white eggs is laid on the bare wood or earth, hatching in 20 days, and about 24 days later the nestlings take flight. Two broods are raised in a season.

Fernbirds spend most of their time well-hidden amongst rushes and shrubs close to the ground, where they build a nest over, or close to, water or swamp.

OFFSHORE ISLANDS

Of all isolated habitats the most solitary are the islands. The sea, whether it be a narrow shallow channel or a vast oceanic expanse is the greatest barrier to the movement of land animals.

New Zealand's 6500 km of coastline is richly endowed with over 500 offshore islands or island groups ranging in size from small rock stacks to several thousand hectares; from white-plumed volcano to subtropical forest. In varying degrees they have all been affected by human activities. Nevertheless, their natural communities have, in general, escaped the gross modification which has occurred on the mainland. As a result these islands have provided havens without which the survival of a considerable number of plant and animal forms peculiar to this country would not have been possible.

During the very early days of colonisation, when it became apparent that many native bird populations were declining because of the increasing destruction of their mainland habitats, the importance of islands as final refuges was appreciated by a few far-sighted people. What also became apparent was that isolation alone would not always guarantee adequate protection for these islands; already farming was being attempted on some, many others were being used as bases for sealing, whaling, and fishing, and alien animals were being introduced both accidentally and deliberately.

About the turn of the century, three important islands, Little Barrier, Kapiti and Resolution, were declared reserves. Since then, many more have been given some protective status to curtail human activity wihout totally prohibiting public access so that the richness and diversity of native life which was once the rule rather than the exception may be glimpsed. On others access is tightly restricted to specialist research workers and management personnel.

On these islands, many natural land communities have been preserved to a far greater extent than elsewhere, often insuring the survival of a species otherwise endangered. This situation is one of great scientific interest and value as, when studying and comparing conditions in these sometimes untouched and sometimes much modified environments, it is possible to evaluate the effects of human influence. The patterns of survival and interaction differ from island to island, each being a dynamic illustration of ecological principles in action, and demonstrate the versatility of nature's role in different circumstances.

Three Kings Islands

Notable mainly from a botanical aspect these islands have an extremely important endemic insect fauna. They are also the only home of the rare, large land-snail, *Placostylus bollonsi*, which, being closely correlated with Pliocene land distribution, evolved in isolation on the islands. Most importantly the islands are free of rodents. One of New Zealand's largest skinks, *Leiolopisma fallai*, is found exclusively on this group of islands. Growing to a length of 30 cm, it is very heavy-bodied with back and sides of various shades of brown, from copper to chocolate, and darker bands down each side. It is a diurnal species fond of basking but in bad weather retreats into holes and under rocks.

Bush birds include the morepork, red- and yellow-crowned parakeets and the small but colourful banded rail. Once widespread in New Zealand this swamp-loving bird is now reported only from coastal areas in north-west Nelson and the northern half of the North Island. It feeds on small molluscs, crabs, insects and worms, as well as a selection of seeds and berries.

The Three Kings also has its own subspecies of the melodious bellbird, differing from the mainland species in a reduction of yellow in its plumage, the undersides being whitish, and overall somewhat larger. Its dependence on nectar is not so evident as with the mainland species, and it has become predominantly insectivorous. Fairly common amongst introduced birds is the Australian brown quail,

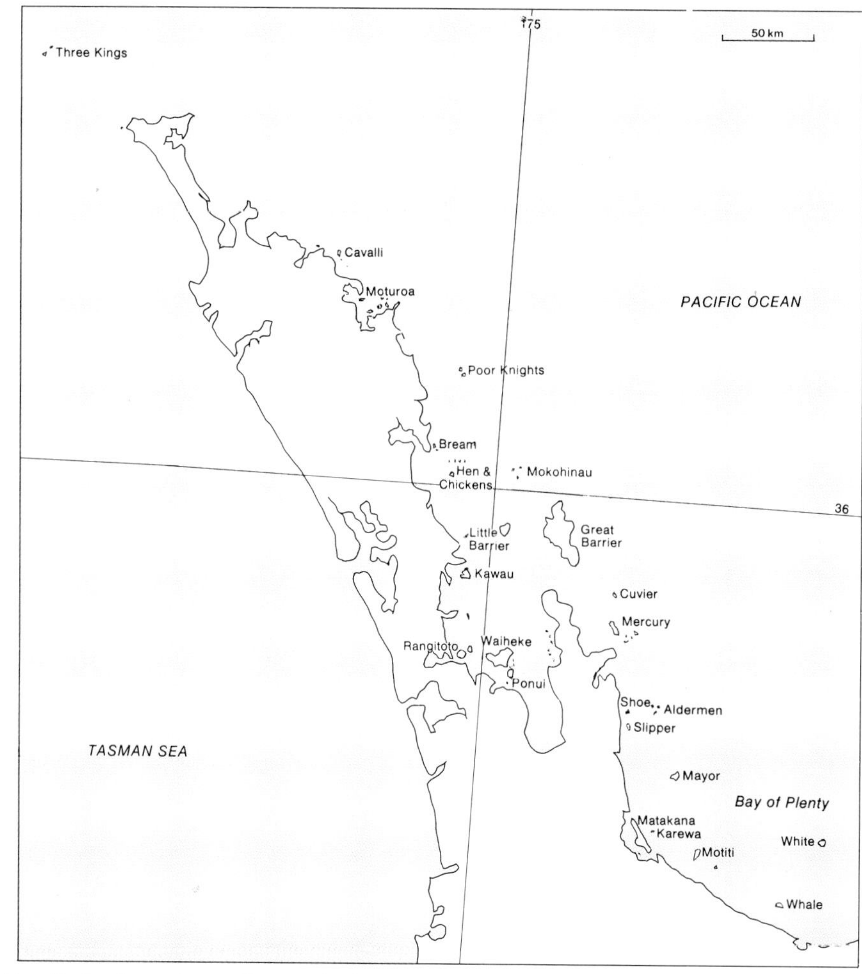

New Zealand's northern offshore islands.

encountered in small coveys, and on being flushed, flies a short distance and settles again.

Seabirds breeding on the island include gannets, grey-faced, white-faced, diving and black-winged petrels; sooty and fluttering shearwaters, and red-billed gulls.

Permission to land must be obtained from the Commissioner of Crown Lands, Auckland.

Poor Knights Islands

Once the site of a volcano and lying 24 km off the coast north-east of Whangarei, the Poor Knights comprise two large islands and several stacks and rocks. Tawhiti-Rahi, the largest and northern-most island, has a moderately flattened top some 240 m above sea level. To the south and separated by a small passage is Aorangi, an island roughly conical in appearance with a large hanging valley facing north-east.

These islands, reserves within the Hauraki Gulf Maritime Park, are bounded by steep bluffs, and cliffs dropping to 90 m below sea level, and are noted for their remarkable tunnels, caves and natural arches. The vegetation is of considerable botanical interest, and includes the rare lily *Xeronema callistemon* which is endemic to the islands, as well as to Hen Island.

The Poor Knights possess a wider variety of animal life than any similar sized group around the coasts. The two main islands carry a good population of tuatara, two species of gecko and five species of skink. It is the only known nesting place of the Buller's shearwater, a breeding ground of the northern fairy prion, and the surrounding seas contain many rare and beautiful forms of marine life including coral, sponges and reef-dwelling fish.

Three forms of the *Placostylus* land snails are found in New Zealand, all restricted to the far

The banded rail's nest of dry, broken rush stems, is usually built in thick swamp, sometimes over water on a thick layer of matted rushes and concealed by surrounding vegetation pulled down to form an arbour.

Falla's skink, one of the largest of the New Zealand skinks, is afforded some protection on the Three Kings Islands.

north of the country. Most abundant of the three is *Placostylus hongii* which evolved in the eastern part of the Northland peninsula and some of the offshore islands, notably the Poor Knights. This peculiar flax snail with a tall spiral shell is primarily a creature of the forest floor, feeding mainly on fallen leaves, those of the karaka tree being especially favoured. The island populations support the belief that salt-laden air appears to be a prerequisite for their survival.

The native *Phormium* flax plant also plays host to an earth-brown weevil which reveals its presence by scalloping the edges of the strap-like leaves. It does this at night, hiding by day amongst dead leaves at ground level.

Most spectacular of the giant wetas, the giant tree weta, *Deinacrida fallai*, is of similar size (73 mm) and habits to the largest species on Little Barrier Island, described later. It is also the only place where the giant cave weta is found. This spectacular insect with a 6 cm body but measuring about 30 cm from the tip of its antennae to the claws on its hind legs (when stretched out), lives in the lava caves and larger crevices. All these sizeable arthropods could only exist in the absence of cats and rats.

Though none of the lizards that occupy the differing habitats these islands afford, are endemic to the Poor Knights, all but two, one gecko and one skink, are restricted to offshore islands. *Leiolopisma oliveri* is a particularly handsome skink, moderately large, with a chunky

A prolific breeder, Buller's shearwater is by far the most common avian species on the Poor Knights. When disturbed it has a raucous scream and a savage bite.

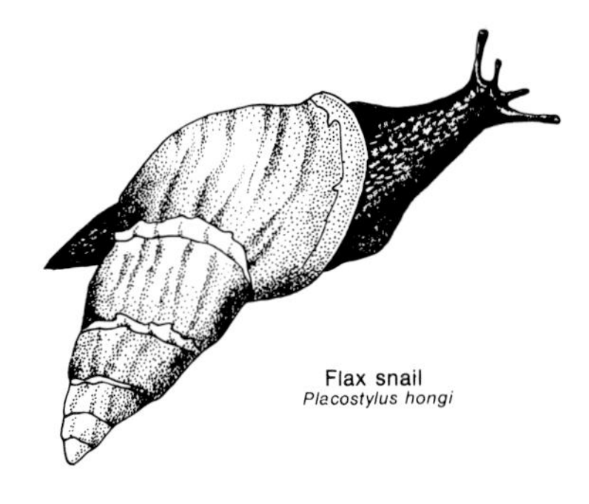

Flax snail. Although the flax snail is a litter-dweller, the juveniles ascend trees soon after hatching but return to the ground when they have grown larger. Below: *The Poor Knights are the remains of lava domes which have long since collapsed under marine erosion, resulting in their isolation, and the consequent preservation of their flora and fauna.*

body, and forages by night in search of *Coprosma* berries and native pepper, moths and beetles.

Also nocturnal is *Hoplodactylus duvauceli*, our largest gecko. This heavy-bodied and strikingly patterned lizard often exceeds 30 cm on the Poor Knights and is, in fact, quite large by world standards. During the day they may hide under stones or bark, in crevices or petrel burrows, and forage at night on the forest floor, frequently climbing high into the trees seeking berries, nectar and insects.

The diurnal lizards of the Poor Knights provide the many kingfishers with a healthy diet, but many of them in turn prey upon small invertebrates such as spiders, insects and crustaceans. The fearsome-looking giant wetas are preyed upon by both lizards and birds but still exist in quite large numbers.

The Poor Knights group supports a host of seabirds apart from large numbers of bellbirds, parakeets, some silvereyes and fantails, and finches in search of winter berries. The most common species is Buller's shearwater, an aggressive coloniser, which was first observed by Buller in 1884. Prior to its discovery on Tawhiti-Rahi in 1914 it was considered a rarity, yet its burrows on the island are now so dense that the ground is perforated with their excavations and walking over them can be hazardous. These islands are the only known breeding ground of this handsome seabird.

Most petrels have mastered the art of climbing trees and those on the Poor Knights have expanded their populations into areas where large trees with rough bark, like pohutukawa, allow ready ascent. The fairy prion, little shearwater and Pycroft's petrel are among those that breed on the islands, also the fluttering shearwater, which, unable to climb like its relatives, is chiefly restricted to the flax lands and nearby cliffs from where they launch themselves into flight. The black-winged petrel is a more recent arrival and a small colony of these attractive seabirds is now nesting on Aorangi.

Permission to go ashore must be obtained from the Commissioner of Crown Lands, Auckland.

The Hen and Chickens Group

Hen Island, or Taranga, is a

Duvaucel's gecko, the largest species of the Hoplodactylus *genus, has no doubt been easy prey for introduced mainland predators. It is now found mainly on offshore islands.*

The Tuatara

Although the tuatara is very like a lizard in external appearances, it is actually the only surviving species of a separate group of reptiles, the Rhynchocephalia, meaning 'beak-headed'. It is found on about thirty islands off the coast of New Zealand where it has probably been restricted for hundreds of years. All of these islands have characteristics which make conditions suitable for these slow growing 'living-fossils'. The Poor Knights Islands typify these features: boulder-strewn beaches, high cliffs, patchy scrub and a forest floor riddled with the burrows of seabirds. Ideal food is plentiful: moths, beetles, crickets and wetas form the major part of their diet and they have also been observed to catch small skinks and geckos.

The tuatara lays its eggs in clutches of 8-15 during late spring in a shallow depression or short burrow. Afterwards the hole is filled with soil and the adult tuatara pays no further attention to it. Before hatching, some 12-15 months later, the embryo develops a horny egg-breaker which is used to slit the soft shell. Emergent young have a very prominent 'third eye' in the middle of the forehead and are coloured with areas of silvery-white on head and tail. Sexual maturity is reached after about 20 years and the sexes are distinguished by the greater size and weight of the male. The lifespan of the tuatara is not known but is estimated to be at least 100 years.

Pycroft's petrel will occasionally share a burrow with the tuatara, although not always amicably!

remarkable sculptured mass of precipitous rock rising steeply from the sea. The Chickens, or Marotiri Islands, consist of eleven islands the largest being Lady Alice, Coppermine and Whatupuke. The entire coastline of the group comprises rough, short boulder beaches alternating with wave platforms and cliffs descending into deep water.

Several species of fauna (and flora) exist here that no longer appear on the mainland, including the tuatara, Pycroft's petrel, three skinks and a gecko, a giant weevil and the rare saddleback. The Chickens are a major breeding ground of the flesh-footed shearwater.

Prior to the conservation programme initiated by the Wildlife Service and others in 1949, the Hen was the last remaining habitat of the North Island saddleback. Since then the success of the programme has ensured the survival of this attractive species and new colonies have been established on Whatupuke, Lady Alice and other islands of the Gulf. The starling-sized saddleback is a member of the family of wattlebirds, glossy with a handsome chestnut saddle on the back. Their legs and bill are also black, but at the base of the bill are bright orange wattles.

Permission to land must be obtained.

Little Barrier Island

Densely forested and rising steeply to a height of 722 m, Little Barrier, or Hauturu, is the largest of the Hauraki Gulf Maritime Park reserves. There still exists an excellent altitudinal sequence of North Island lowland forest undisturbed by grazing animals and with only a small area modified by man. The uniqueness of the island has long been recognised by naturalists, and despite the possible presence of wild cats and Polynesian rats the abundant bird life continues to thrive.

The Polynesian rat, or kiore, is the smallest of the three species in New Zealand. Its fur is quite long and silky, brown on the back and greyish underneath. In pre-European times, kiore were apparently common through much

The North Island saddleback is probably a descendant of a race considered to have been a native since before the time, at least 80 million years, when New Zealand first became an archipelago.

Kiore, a native of South-East Asia, is widely distributed in the Pacific where it is believed to have been carried in the canoes of the migrating Polynesian peoples from Asia 1500-2000 years ago.

of our lowland forest and were a favoured food of the Maori. However, after the arrival of European rats and mice, kiore quickly disappeared from most of their former habitats. By 1850 they were extinct in the North Island except for a number of offshore islands.

A local species of giant weta of the genus *Deinacrida* is one of the most spectacular of the wingless insects. Reaching a body length of over 9 cm and weighing up to 80 g, it is the heaviest insect endemic to New Zealand, and one of the largest in the world. Rarely seen by day but emerging after dark to feed on the foliage of trees, this species was known to the Maoris as weta-punga; punga was a Maori god who ruled over deformed and ugly creatures. The flying insects, as on other islands not too far offshore are the same species as those on the mainland.

Ten indigenous species of earthworms occur on Little Barrier, two of which are found nowhere else in New Zealand. One, *Spenceriella gigantea*, is our largest worm. It is a subsoil species, attaining a length of 130-140 cm. Their deep burrows are made principally not for shelter, but to obtain food by the ingestion of soil. Only occasionally do they come to the surface. This giant worm is also found on islands in the Hen and Chickens group.

Five skinks and five geckos are known to inhabit Little Barrier Island, including the only New Zealand lizard to lay eggs, the nocturnal shore skink. This species is quite at home in water which is even more unusual and is sometimes seen swimming in brackish pools presumably feeding on aquatic larvae.

All the geckos present on Little Barrier, with the exception of Duvaucel's gecko, which is restricted to these northern offshore islands and Cook Strait, are reasonably common and very widely distributed throughout the North Island.

Among a prolific avifauna, the stitchbird or hihi is particularly important because it is unique to this area. Once, the stitchbird was found in dense bush throughout the North Island but has not been recorded on the mainland since 1885. The male is a brightly coloured bird with a velvety-black head, conspicuous white ear tufts and a broad band of

Today the giant weta survives only in small numbers on Little Barrier Island. Wetas are thought to be representatives of a very ancient group of insects which have changed little since they originated. They have a long fossil history dating back to 190 million years ago.

An adult male stitchbird on Little Barrier Island, which is the only remaining habitat of this honey-eating species.

yellow across the breast. The female is much duller, more closely resembling the female bellbird both in size and colour. Stitchbirds are easily identified in silhouette by their erratic movements and often tilted tail, and by their staccato 'tzit,tzit,tzit' call.

The stitchbird is not the only rare member of the island's birdlife. The black petrel or taiko, though found off many parts of our coastline, nests only on the Barrier Islands, and even its presence on Little Barrier is not as secure as it once was: cats and rats have taken their toll. Closely related is Cook's petrel – very similar in appearance to Pycroft's petrel, but distinguished by its longer bill. Little Barrier Island is the main breeding centre for Cook's petrel and large numbers may be found in the summer months on the

The strongly hooked bill of Cook's petrel serves as both efficient tool to dig burrows in which to lay its eggs, and as effective weapon against intruders.

The Peripatus

The fallen leaves that carpet the forest floor conceal a diversity and complexity of living things almost unimaginable. Most of this life is small and nearly exclusively invertebrate. Besides providing an abundance of food, the forest floor also maintains a haven for many kinds of creature that simply could not exist elsewhere. The herbivores feeding on plant material in turn provide food for the carnivores which hunt them. Most important among predators are arachnids: spiders, harvestmen and mites. Least seen among the predators is a velvety-blue caterpillar-like animal called the peripatus. Belonging to the phylum Onychophora, this curious creature is not a caterpillar, neither is it a worm, but has certain characteristics of both insects and worms. Despite its soft harmless appearance, it is an active predator, feeding mainly on insects, and is often found eating already dead invertebrates including its own kind. When disturbed it can defend itself very effectively by shooting out a stream of sticky fluid from paired glands near its mouth.

Although the peripatus must have been one of the earliest animals to adapt to a terrestrial environment, it is still found only on the floor of wet forests because it has no way of controlling the loss of moisture through its skin. The species alive today, found throughout countries of the Southern Hemisphere, are specialised relics little different from ancestors of 500 million years ago.

The most common of New Zealand's representatives in this phylum is *Peripatoides novae – zealandiae*, which grows to 35 mm in length, and has fifteen pairs of stumpy legs. It is a cautious, slow-moving creature with little sense of sight or smell, relying primarily on a very acute sense of touch. The young are born alive, but another smaller species, *Ooperipatus viridimaculata*, common in the leaf litter of the far south, actually lays eggs.

higher ground. Their burrows are about 12 cm in diameter at the entrance and can be up to 4 m in length. The only other known breeding place is Codfish Island in Foveaux Strait, where they are under threat from predation by wekas.

Bush birds of Little Barrier are many and include the native pigeon, tomtit, bellbird and tui, rifleman, kaka, robin and fantail. The North Island brown kiwi introduced from the mainland is common, even on the highest ridges. The long-tailed cuckoo is abundant as a summer visitor.

There are also some tuatara although numbers are declining, as well as colonies of short-tailed and long-tailed bats. The yellow admiral butterfly is often seen at the summit.

Mokohinau Islands

Equidistant between the Hen and Chickens and the northern tip of Great Barrier Island lie the Mokohinau Islands. The largest of the group, Burgess Island, is an automated lighthouse station. The remainder, Fanal, Flax, Trig and Groper Islands together with a number of smaller rocks and reefs are also within the Hauraki Gulf Maritime Park. The islands are bleak, vegetation is stunted and fauna is comparatively sparse. A number of lizards are present and birdlife includes gulls, muttonbirds, pigeons, shags and red-crowned parakeets. There is a large breeding colony of red-billed gulls and a small colony of diving petrels.

Permission to land must be obtained.

The red-crowned parakeet is now very rare on the mainland but continues to thrive on many offshore islands, its preferred habitat being open country at forest margins.

Kawau Island

A popular resort 50 km north of Auckland, steeped in Maori and European colonial history, and made famous by the eccentric Sir George Grey. In 1862 Grey purchased Kawau and the mine manager's house (copper was mined in a small way) was converted into the imposing Mansion House. To the native bush were added exotic trees and shrubs from many parts of the world; then came the animals. A pair of zebras died soon after their arrival; monkeys did so well they had to be exterminated as pests, but some deer, opossums, and, above all, the wallabies, flourished.

Of the twelve marsupial species liberated in New Zealand before 1870, only the brush-tailed opossum and six species of wallaby became established. One species, the red-necked or brush wallaby was let loose in the South Island: all the others are found on Kawau, with the exception of the black-striped wallaby which is thought to have died out.

Most numerous of the Kawau wallabies is the tammar, or dama wallaby, which favours the bush and clearings in the south and centre of the island. Differing habitats suit different species. Parma wallabies prefer the thick manuka scrub in the north, whereas the black-tailed or swamp species survive in damp areas. The rock wallaby, the most colourful of all with its coppery brown coat, dark bushy tail and dark markings on the face, is confined to the cliffs and rocky slopes. This species is also present on Rangitoto

Above: *Swamp wallaby with young in her pouch. Sometimes the female conceives just after giving birth and the suckling of the joey interrupts the development of the embryo which lies dormant, but very much alive, until suckling ceases.* Above right: *The dama wallaby lives in thick cover and can survive on very dry food.*

Island and Motutapu.

The wallabies are all grazers, and some, like the swamp wallaby, also browse amongst shrubs and trees. They are mainly nocturnal, spending most of the day hidden in dense cover, coming out to feed an hour or so before dusk.

Like the opossums, the female wallabies carry their young in a pouch, or marsupium. At birth the baby is extremely small, not much larger than a bean, and yet is able to find its way, unassisted, to the pouch, where it firmly attaches itself to a nipple. There it remains for many months, entirely dependent on its mother until its own short forays gradually lead to independence, sometimes up to a year after birth.

The kookaburra was introduced between 1860 and 1880 but the descendants now survive only north of Auckland among the creeks and islets along the western shore of the Hauraki Gulf.

Several bird species are to be found in the variety of habitats that make Kawau Island such an interesting place. Again, Governor Grey was responsible for the introduction of a many of them, of which the kookaburra is a conspicuous example. In Australia they are well known as snake-killers but here, in the absence of snakes, their substantial diet consists of lizards, eels, crabs, wetas and the occasional small bird. These large forest kingfishers can be heard at dawn and dusk when their demonic laughter shatters the monotonous chorus of the insect world. Hence their alternative name of laughing jackass. The vividly coloured Australian rosellas are occasionally seen here and seem to be increasing on the mainland also. Around Kawau are several small islands where modification is minimal, seabirds breed and shags dry their wings.

Goat Island

Situated 25 km north-east of Warkworth this small island is a rare combination of open and sheltered coast with varying depths of water, isolated but reasonably accessible – an ideal situation for a marine laboratory.

On the south-western end of the island there is a small colony of red-billed gulls and some white-fronted terns. Nearby are the burrows of a small number of flesh-footed shearwaters. Pied shags inhabit the pohutukawas along the shore, while the occasional gannet patrols the surrounding waters. Blue penguins and diving petrels are frequent visitors, both adept underwater swimmers.

Rangitoto Island

It is approximately 200 years since the last eruption of the largest and youngest of Auckland's many vol-

The white-fronted tern – a gregarious species that breeds only in New Zealand.

Precipitous cliffs of Kawau Island favoured by the rock wallaby.

canoes, Rangitoto (literally 'blood-red sky' implying dramatic activity during early Maori occupancy), the island whose characteristic outline is the same from most parts of the mainland. The scoria crust of this almost circular island contains no soil, but the rich volcanic dust and humus gradually accumulating in the fissures supports an incredible variety of indigenous flora, including over 40 species of fern, a score of orchids, 100 or so mosses and liverworts, and a canopy of pohutukawas and shrubs established from windblown seeds and berries dropped by birds.

Opossums, wallabies and fallow deer were released on the neighbouring island of Motutapu about 1870, and crossed by way of a natural causeway linking the two islands. A rapid increase in their populations caused serious damage to the vegetation and eradication programmes have been necessary to control their numbers. Today, the 60 cm tall rock wallaby is still present in quite considerable numbers, living and feeding in grassy areas above the cliffs. It is a very agile creature and can jump as high as four metres onto sloping trees.

The skink *Leiolopisma smithi* lives under piles of driftwood, seaweed and stones above the high water mark, feeding on the numerous forms of invertebrate life found in such situations. It is a diurnal species and may be found basking on exposed rocks and boulders.

Sparrows and finches are common, as are blackbirds, thrushes and gulls. Of special interest on the western lava field is Auckland's largest colony of black-backed gulls.

Waiheke Island

Waiheke is the largest island within the Hauraki Gulf, and the most developed. Several residential settlements are scattered around the numerous sheltered bays and most of the interior is open pasture where sheep-farming is the predominant industry. Despite the development on the island, Waiheke performs an important role in the Hauraki Gulf's ecology providing shelter for seabirds and shallower waters for the variety of intertidal marine life.

The New Zealand dotterel breeds

Heavy-bodied, coppery brown on the back and sides with distinctive alternating light and dark-brown back markings, Leiolopisma homalonotum *is one of the handsomest of the New Zealand skinks.*

on open sandy beaches on Waiheke, Ponui and Motuihe Islands. Small colonies of spotted shags inhabit Waiheke as well as rocks in The Noises group.

Great Barrier Island

Largest of the North Island's offshore islands, Great Barrier is a detached remnant of the Coromandel Peninsula, which it closely resembles in structure, but not quite matching the latter in its diversity of flora and fauna. The birds are numerous around the coast but fewer in the hills. Goats and pigs are scattered, rabbits widespread, and mustelids absent.

Five species of skink and four of gecko live on the island. *Leiolopisma homalonotum*, one of New Zealands rarest skinks, is now found only on Great Barrier Island, where it lives on the forest floor. It shelters under logs and stones and has been seen basking in shafts of sunlight deep in the forest.

Two species of rat are present: the ship rat and the kiore, both of which are hunted by a small population of feral cats. Despite these predators, the colony of black petrels in the forests around Mt Hobson appear to be flourishing. Black and pied shags are numerous and a gannet colony occupies part of an island off the west coast.

Few introduced birds have made an impression on Great Barrier. The swallow has arrived but the sparrow has not. Kingfishers are everywhere and the shining cuckoo is common in the summer months, especially in association with grey warblers, chaffinches and whiteheads. The shining cuckoo is known to breed only in New Zealand, arriving in August and September from the Solomon Islands, and making the return journey in February and March. Its diet which consists almost entirely of insects and their larvae, also includes the hairy caterpillar of the magpie moth, conspicuously avoided by all other birds. It is a small sparrow-sized bird, the upper surface having a copper glint on a metallic green. Underparts are off-white, and barred with the same shining colours. It is a bird more often heard than seen from North Cape to Stewart Island.

The harrier is abundant, the North Island kaka is holding its own and the kokako is present but very secretive. Smaller bush birds, including the fantail and silvereye are plentiful. The island is the principal remaining sanctuary of the rare brown duck.

The migratory shining cuckoo is an habitual spring and summer visitor to many forest areas of New Zealand. Unlike migratory seabirds, cuckoos must remain airborne in the absence of land. Their non-stop flight, therefore, could be up to 3000 km.

The pied shag is basically a salt-water shag, though some colonies occur a short way inland over fresh water. Their distribution is uneven since they avoid long exposed coasts.

Cuvier Island

This island has a dual role on the edge of the Hauraki Gulf. Twenty-four hectares are in the lighthouse reserve and the remaining 171 ha is a nature reserve within the Hauraki Gulf Maritime Park. Since the eradication of goats and feral cats in 1960 the native bush has rapidly regenerated and saddlebacks, introduced from Hen Island, have thrived where old pohutukawa trees provide nesting holes. Tuataras are few, sharing the burrows of petrels and shearwaters.

The insects are a mixture of bush and agricultural species, the lighthouse reserve being largely in pasture. The ground beetles are conspicuously few, perhaps owing to a very large population of kiore. A large darkling beetle, an important part of the tuataras' diet on islands where they coexist, reaches its northernmost distribution on Cuvier Island.

Permission to land must be obtained from the Commissioner of Crown Lands.

Mercury Islands

These islands are particularly interesting to ecologists for the study of the effects of the kiore. Where there are large numbers of kiore, there appear to be small numbers of tuatara, lizards and certain bird species. Middle Island in the Mercury group is one of the few islands unmodified and free of rodents, and subsequently supports a larger and more diverse population of invertebrates and birds than the other rat-inhabited islands of the group. A rare weta – only one specimen has ever been found – is thought to live on this island. Not yet scientifically described or named, this 5 cm insect, unlike its relatives elsewhere, sports two large horns that point aggressively forward from the sides of its head.

Seabirds that breed on the Mercury Islands include Pycroft's and grey-faced petrel and the white-faced storm petrel and the flesh-footed shearwater. Vast numbers of diving petrels nest on Middle and Green Islands. Pied shags colonise the pohutukawas that clothe the shore; brown quail and pukeko are abundant in low scrub and wet pastures respectively; small bush birds and introduced finches are fairly common, red-crowned parakeets add a dash of colour and wood pigeons glide in the open. Apart from visiting starlings, the skylark is probably the most common bird on Great Mercury. Saddlebacks were released on Red Mercury in 1966.

Large populations of spiders and insects that prefer salt-laden breezes inhabit the boulder beaches and coastal cliffs of Red Mercury in particular. Spiders of the forest and scrub are similar to mainland species, though trap-door, wolf and crab spiders are absent. Green and Middle Islands have the largest populations of lizards in the group.

The white-faced storm petrel lives at sea for 5 months of the year, remaining for 7 months on or near breeding grounds.

With the exception of Great Mercury, which is private, all the islands in this group are Nature Reserves and permission to land must be given by Crown Commissioner of Lands, Auckland.

The Aldermen Islands

This group, gifted to the Crown by the Maori owners for nature conservation during a visit of Queen Elizabeth II, comprises three main islands in a triangle with a chain of pinnacle rocks across the centre. It was named 'The Court of Aldermen' by Captain Cook, no doubt because of its imposing appearance and formation. The vegetation is prolific in sheltered areas, while in exposed places it is scrubby, but supports a considerable seabird population and large numbers of tuatara. Grey-faced petrels are abundant and Flat Island is the home of a large colony of white-faced storm petrels, sometimes known as 'Mother Carey's chickens.' There are also numerous bellbirds and parakeets.

Permission to land must be obtained.

Leiolopisma pachysomaticum *is a medium sized but heavy-bodied skink with an irregularly speckled mid-brown dorsal surface and a cream or grey under surface, densly spotted with dark brown.*

The Shoe and Slipper Group

Lying just off the Coromandel shore south west of the Aldermen, this group is privately owned, Shoe being under Maori ownership. Here too, native birds flourish and the blue penguin probably breeds in small numbers. Seventeen species of land snail have been recorded on Shoe Island, all specimens taken from the leaf mould under a canopy of kanuka and pohutukawa. Kiore are present on Rabbit and Penguin Islands, the disreputable Norway rat on Slipper. Skinks are found on all three, but the apparent absence of the black shore skink suggests that Shoe Island and the Aldermen, where this lizard is common, form its southern geographical limit. One skink, *Leiolopisma pachysomaticum*, is unique to the Aldermen Islands. Medium sized and squarish in body section, it is a forager of the leaf litter below the coastal forest and although frequently active during the day has also been observed to fossick at night.

Mayor Island

A hilly bush-clad volcanic island in the west of the Bay of Plenty with rocky cliff-bound coast fringed with pohutukawas. The centre of the island is occupied by a crater, 8 km in circumference, in which there are

The male and newly arrived female gannet greet each other with outstretched wings, raised heads and excited calls.

two lakes – one a deep blue-black, the other blood-red from a distance, the colour being caused by the algal growth. These below sea-level lakes are of particular interest because of the absence of eels and freshwater crayfish, but small colonies of shags and scaup nest in the vicinity and kaka take refuge in dense bush.

White Island

A yachtsman's description of White Island would include adjectives like, ravaged, inhospitable, pungent and dramatic. Constantly emitting steam charged with sulphur dioxide and hydrogen chloride, the active volcano-island is a desolate but spectacular sight. From time to time ash eruptions occur, sometimes explosively. Yet despite the dangers the bird-life is prolific. Boasting the largest gannet colony in New Zealand, the island is a favourite breeding place of red-billed gulls and grey-faced petrels. An 81 ha forest of pohutukawas softens the harsh yellow-grey of the crater's flanks and provides cover for a few passerines.

This island is privately owned and permission to land is required.

Karewa Island

A 4 ha Wildlife Sanctuary in the Bay of Plenty and home of the tuatara, diving and flesh-footed shearwaters and other petrels. Reef herons and blue penguins also breed on this rugged island. Landing is forbidden.

Whale Island

Heavily modified by man, having been occupied by Maoris, used as a

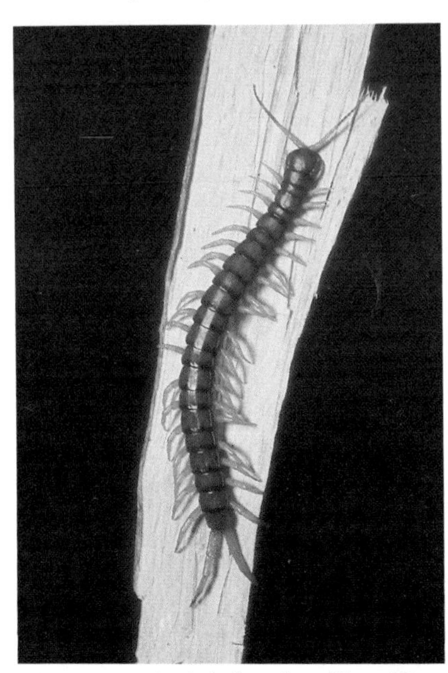

The giant centipede is found on Three Kings and Poor Knights Islands, and the long-legged species is also found on the Hen and Chicken Islands.

White Island, lying 50 km offshore from Whakatane in the Bay of Plenty, is one of New Zealand's three most active volcanoes. Much of the broad crater floor is only a few metres above sea level.

The little spotted kiwi fossicking for food. The animal component of the kiwi's diet, which includes earthworms, centipedes, slugs and snails and some spiders, is balanced with leaves of some native shrubs and grasses and with seeds and berries from a variety of native plants including the nikau palm and kahikatea.

whaling station and quarried for rock, the island is now privately owned but gazetted a wildlife refuge in 1965. It lies on the belt of thermal activity and a few hot-water springs and mud pools are present in the northern part of the island.

Goats were present until 1975 and rabbits, despite eradication campaigns are still widespread. Norway rats are the only rodents but evidence suggests they are to be found wherever there is sufficient cover, from dune areas and boulder banks along the shores to the grassy slopes and bush of the summit. Grey-faced petrels are distributed widely over the island with concentrations in the north, the total population estimated at 40,000 pairs. Sooty shearwaters nest in some parts but their numbers are low because rats are almost certainly responsible for the lack of native snails, wetas and other small animals. The most common native bird on the island is the ubiquitous fantail. There are many blackbirds and the redpoll is a frequent visitor. Four species of skink and two geckos are reported though the rats have taken their toll of these also.

Kapiti Island

Since 1897, when Kapiti Island was declared a reserve, there has been a massive regeneration of native bush. Goats and cats were eliminated by 1940 and a remarkable variety of natural habitats have established since. At present Kapiti Island is the largest single area (165 ha) of lowland and coastal forest in New Zealand that is free of all large introduced mammals. There are no stoats, cats, rabbits, deer, pigs, goats or ship rats. In the absence of these predators Kapiti has gained a unique place in the natural history of New Zealand. Nowhere else with the possible exception of Little Barrier Island, can kakas, parakeets, tuis, bellbirds, pigeons, whiteheads, robins and wekas be seen so easily in the wild.

Many years ago the little spotted kiwi was widely distributed on both main islands. Today 'the little grey ghost' survives on only two islands: D'Urville, where the population is all but extinct and Kapiti where a population of 500-600 birds appears to be in a healthy state. The little spotted kiwi is the smallest of all, averaging about half the weight of the brown kiwi. As in all kiwis the females are bigger and heavier than the males. Both the little spotted and the North Island brown are found on Kapiti, where dense damp forest provides cover and soft mossy ground with rotting leaves and logs providing a variety of worms and insects. They remain concealed during the daytime but at night become very active, fossicking for food with their specially adapted and extremely sensitive bills.

The wekas on Kapiti are a hybrid species descended from a Stewart Island male and a North Island female that were liberated before the island became a sanctuary. They are generally like the North Island species but with redder legs. Other native birds living and breeding on the island include: yellow-crowned and red-crowned parakeets, grey duck, paradise duck, reef heron, black shag, little blue penguin and sooty shearwater. The shining cuckoo and the long-tailed cuckoo are both present, the latter being the more plentiful. Saddlebacks were released on the island in 1980.

Introduced birds breeding on the island include mallard, black swan, California quail and several finches and song-birds. The island's shallow tumbling streams are the haunt of the native trout, eels and freshwater crayfish, while the rotting forest litter is home to a myriad tiny insects. Decaying branches conceal huhu grubs, beetles and wetas, fought over by raucous kakas, while wekas, opossums and kiore compete for fallen berries.

Kapiti Island is often the landfall of strangers blown across the Tasman by strong westerly winds, sometimes to die, sometimes to exist briefly like the blue moon and painted lady butterflies.

Permission to land on this native reserve must be obtained from the Commissioner of Crown Lands, Wellington.

Mana Island

Fifteen kilometres south of Kapiti, Mana Island, though greatly modified by man, is one of the few islands totally devoid of rats, thus many of the invertebrates have been allowed to evolve in secure isolation. Mice, however, do exist. A weta found on Mana differs from

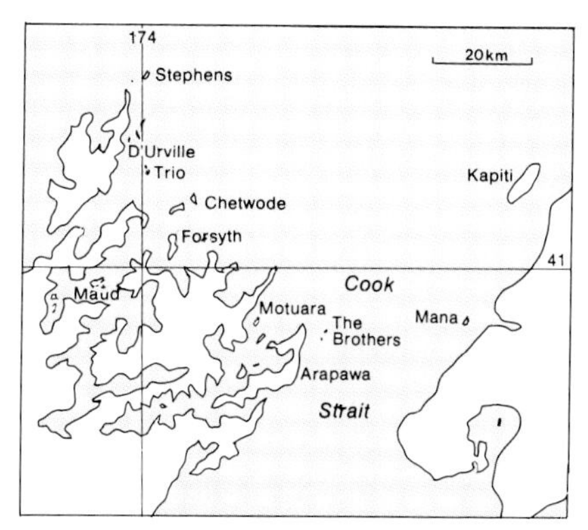

Offshore islands in Cook Strait.

Hamilton's frog, the rarest native frog, is a tiny creature (sometimes measuring only 4 cm) restricted in distribution. Native frogs were unknown to the Maoris yet had existed for millions of years.

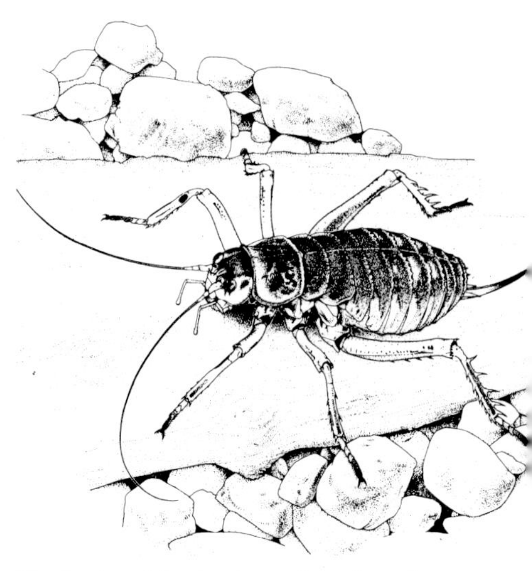

The Stephens Island weta, a subspecies unique to this area.

those found on the mainland, and a skink, known only from two other small islands of the north-east coast, is also resident. It appears as though the present distribution of *Leiolopisma macgregori* represents the two isolated extremities of its range, the mainland population having been eliminated by predators. It is now one of our rarest lizards, living amongst the leaf litter of coastal forest and sheltering during the day under rocks and logs often in the vicinity of seabirds.

Permission to visit is required from the Commissioner of Crown Lands, Wellington.

The Marlborough Sounds Maritime Park

Gazetted in 1973 this Park covers an area of 29,400 ha, from Croiselles Harbour in the west to Port Underwood in the east. It is not a continuous park but contains over 100 separate reserves scattered among the peninsulas and islands of the Marlborough Sounds. The emphasis is on public use, hence some reserves are recreational, others of historic value and some used as sanctuaries for endangered plants and animals. A permit is required for landing on some of the islands.

Stephens Island

This island is one of the more famed around our coasts. Not only is it a home of tuatara but one of only two localities (the other being Maud Island, 40 km south) which harbours what is probably the world's rarest frog. Hamilton's frog was discovered in 1915 among the moss-covered rocks of a boulder bank near the summit of Stephens Island. Only occasional specimens were found in subsequent years and none at all between 1942 and 1950 giving rise to the fear that the species had been lost. The forest covering the island was destroyed by grazing, cutting and burning, exposing it to the salt-laden winds which continually blow through Cook Strait. Then in 1950 the frog was rediscovered, arousing considerable interest in protection of the species. The Wildlife Service commenced the task of re-establishing the vegetation and ensuring maximum protection for a precarious population.

The raucous kakas of Kapiti Island break open rotten branches in their quest for grubs, wetas or beetles.

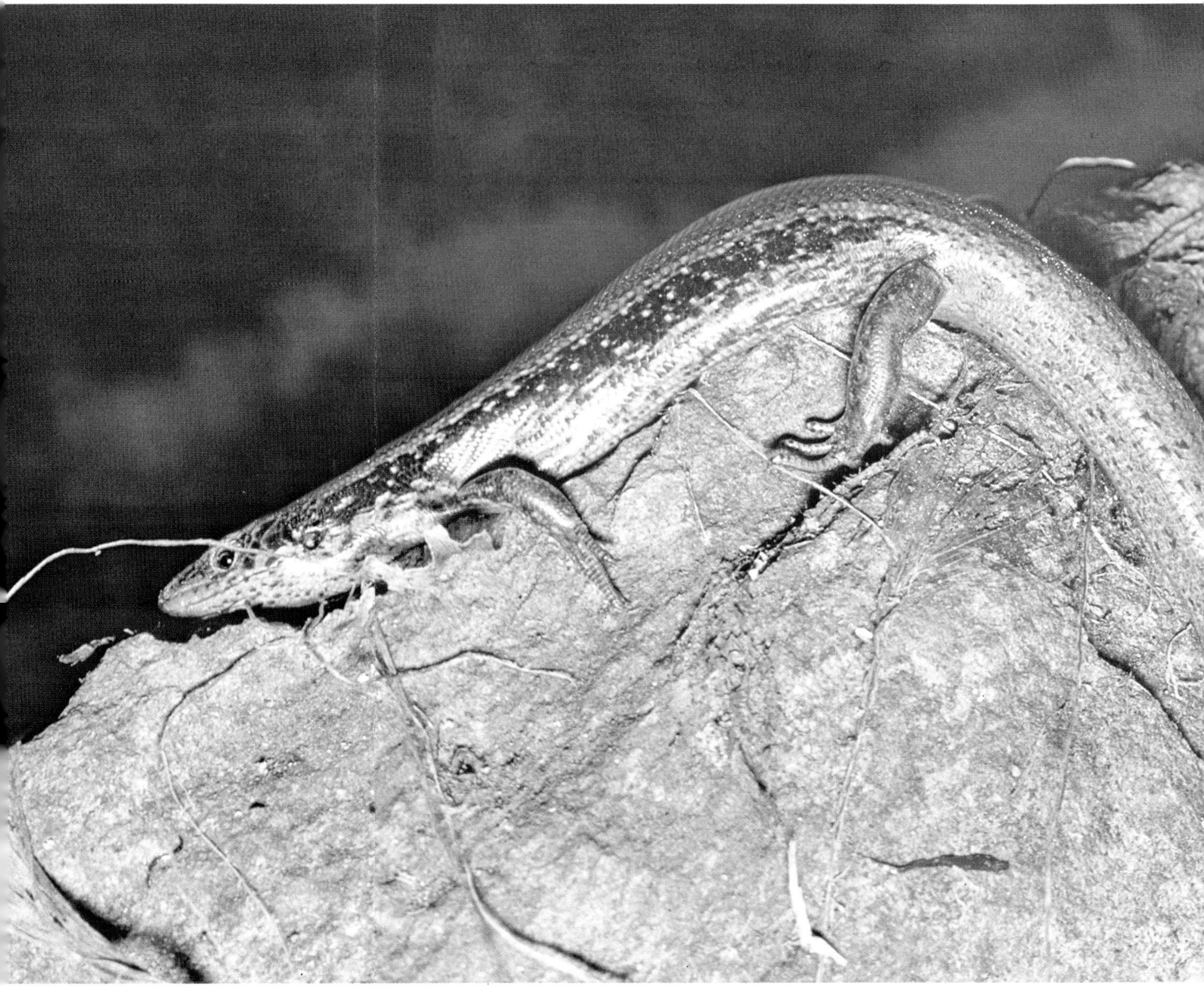

Leiolopisma infrapunctatum, *a moderately heavy-bodied skink of the Cook Strait region.*

In 1958 a colony was found on Maud Island in Pelorus Sound, where it lives in a 15 ha forest remnant. The constant dampness of this habitat is more conducive to its survival. Although our native frogs may not need pools and streams, they must have a moist environment, otherwise their porous skins dry out and they die. On Stephens Island the rocks of the boulder bank trap moisture and dew and also provide some protection from the predatory tuataras.

There are no rats or mice on Stephens Island and this is the single most important reason why the tuatara still thrives. All islands where the tuatara is declining appear to have a total absence of juveniles despite its prolific breeding habit, and all except one have populations of kiore. The conclusions are obvious. Nevertheless this 'living fossil' is not in any immediate danger of extinction, though it was once much more widely distributed than it is today.

All the tuatara islands have populations – some immense – of breeding petrels, the main occupants and architects of the burrows characteristic of tuatara habitat. This association continues to baffle naturalists as the birds undoubtedly provide the tuataras with many benefits, principally in the way of housing and food; but it is hard to imagine how the birds benefit when their eggs or chicks may be eaten!

The diet of the tuatara also includes lizards, frogs and a variety of invertebrates: crickets, moths, spiders, wetas and beetles. Hamilton's frog too, is carnivorous, so between them they must account for a high mortality rate amongst the island's insect fauna.

Frogs and tuatara 'hunt' mainly at night, depending mainly on sight rather than on smell or hearing. Their methods are unusual in that they stay quite motionless and wait for their prey to come to them – ready to snap up anything of a suitable size that happens to wander close by. To exist in such large numbers on Stephens Island there has to be an abundance of food. And there is. Sampling has shown that there is an extraordinarily large insect fauna, with several species endemic to the island, and many which have been listed as endangered or rare invertebrates. They include a large, agressive-looking predatory carabid beetle that burrows in the soil and is endemic to the island *(Mecodema costellum)*, a large weevil, and the Stephens Island giant weta. Another weta that frequents the native foliage, along with a heavy, black darkling beetle, are favoured foods of the tuataras.

Smaller insects and grubs are the staple diet of two geckos and at least four skinks which are known to inhabit Stephens Island. The Marlborough green gecko is an attractive pale green with varying amounts of yellow spots or lines, found mainly in the manuka canopy, and confined to the Marlborough Sounds area. The other gecko is, like others of the genus, strictly nocturnal, and is known only from Stephens Island.

Of the skinks none is more attractive than *Leiolopisma infrapunctatum*. It is a fairly large lizard, beautifully marked and reflecting a dark green sheen in sunlight. Though found in several mainland localities in the vicinity of Cook Strait none are quite so large as the Stephens Island specimens. It prefers relatively open areas close to bush, often in association with the burrows of fairy prions. It is fond of basking but moves very quickly to cover when disturbed.

Permission to land is required from the Wildlife Service.

Trio Islands

These are the least modified of all the Cook Strait islands supporting vast numbers of diving petrels, fairy prions and fluttering shearwaters. Tuataras often share the burrows, and lizards are frequently found in association.

Under Maori ownership, the island is private and permission to land must be obtained from the Internal Affairs Department.

Maud Island

This island in Pelorus Sound was once wholly forested but now all but 30 ha has been cleared for grazing. Affording less severe conditions than Stephens Island it is a haven for Hamilton's frog which exists in greater numbers. Maud is totally free of mammalian predators and the Wildlife Service has successfully introduced the South Island saddleback and the kakapo in recent years.

The colourful saddleback, the size of a starling and found only in New Zealand, is a member of the ancient wattlebird family, one relative of which, the huia, is presumed extinct; its other member is the kokako or native crow. The South Island saddleback is very similar to its North Island cousin in appearance and habit, only lacking a pale-yellow band that separates the 'saddle' from the glossy black body feathers. Saddlebacks are strongly territorial and once a pair is established they will defend their territories by using a series of loud, long-distance calls. They are active almost continuously during the day in search of food which consists of insects and a variety of native fruits.

Permission to visit is required from Department of Internal Affairs and the Commissioner of Crown Lands, Blenheim.

D'Urville Island

The third largest of New Zealand's offshore islands after Stewart and Great Barrier, it is greatly modified and exhibits very similar faunal characteristics to mainland open country. A small population of little spotted kiwis have a foothold on D'Urville.

Chetwode Islands

This is a small group of islands standing across the entrance to Pelorus Sound. The two larger islands are Te Kakaho and Nukuwaiata with several rocks and stacks close by, the most prominent of which is Sentinel Rock to the north. Nukuwaiata is clothed in mature native forest and provides a similar habitat to Kapiti Island. Bird life is abundant and varied. Both islands maintain South Island robins and yellow-crowned parakeets in significant numbers. Te Kakaho supports a nesting colony of fluttering shearwaters.

Permission to land is required from the Commissioner of Crown Lands, Blenheim.

The Brothers

Little more than a cluster of rocks emerging from the cold waters of Cook Strait, this group is inhabited by large numbers of fairy prions, petrels and, sharing their burrows, the tuatara. The fairy prion is commonly seen in offshore waters and further at sea flying just above the crest of the waves. At a distance

Unlike many of their oceanic kin, diving petrels prefer inshore waters where they dive freely in search of small fish and planktonic crustaceans. These birds probably originated from near Cape Horn.

South Island saddleback. While numbers have decreased as a result of European-induced changes and the introduction of rats, the saddleback is thought to have already been in decline before European settlement.

The King Shag

An impressive looking bird, the king shag appears from a distance to be black and white, but the light playing on its plumage reflects a steely-blue sheen and on its wings, oily-green. The undersurfaces are white and there are white patches on the wings and back. Its feet are flesh-pink and the eyelids a conspicuous blue. The king shag is the largest of the subantarctic shags, but slightly smaller than the black shag which is as much at home far inland as it is along the coast. The king shag is also the rarest: the total world population is restricted to four breeding colonies in the Marlborough Sounds – White Rocks, North Trio, Duffers Rock and Sentinel Rock, and landing is strictly prohibited.

The breeding season extends over the winter months, but the birds do not move far outside the breeding area. They feed on bottom-living fish such as sole and sand eels, sometimes including red cod and crabs. Two eggs are laid in an untidy nest built of seaweed on exposed rocky platforms. When it is hatched the chick is black and naked, but soon grows a coat of sooty down and is ready to fly at six weeks.

The king shags of Sentinel Rock share their habitat with white-faced storm petrels – the only breeding colony in the Cook Strait region.

The fairy prion, one of the breeding species of seabird common in the Hauraki Gulf. It is also found in the Snares group.

it is rather inconspicuous on account of its dove-grey plumage but when turning in flight exposes its undersurface, and at close quarters it may be recognised by the black W across the outstretched wings and a dark band on the tail.

Permission to land is required from the Department of Internal Affairs.

The Islands of Fiordland

The fiords of the south-west coast are littered with islands large and small. The largest – Resolution Island – was proclaimed New Zealand's first bird sanctuary in 1894, and though a certain Richard Henry, ranger, spent years transferring endangered species to the island, it is probable that the presence of the stoat has exterminated the kakapo and kiwi populations.

Most of the Fiordland region remains unchanged since Cook and other early adventurers gazed upon the primeval scene two centuries ago. In an attempt to preserve it thus some areas have been designated 'Wilderness Areas', where special efforts are made to prevent the introduction of exotic animals and plants. One such place is Secretary Island, where access has been very restricted since 1963. Rats and stoats have probably managed to swim there from the mainland but there are no opossums or deer. Habitats consist of lowland and upland forest, subalpine scrub and grassland with no modification. No kakapo or takahe have been reported but kiwis and wekas are plentiful and there is a limited variety of smaller birds.

Two species of penguin are usually present among the rocks and islands and mainland coasts of Fiordland; the one most likely to be encountered being the Fiordland crested penguin. The prominent yellow eyebrows are a distinctive feature of this handsome bird: its

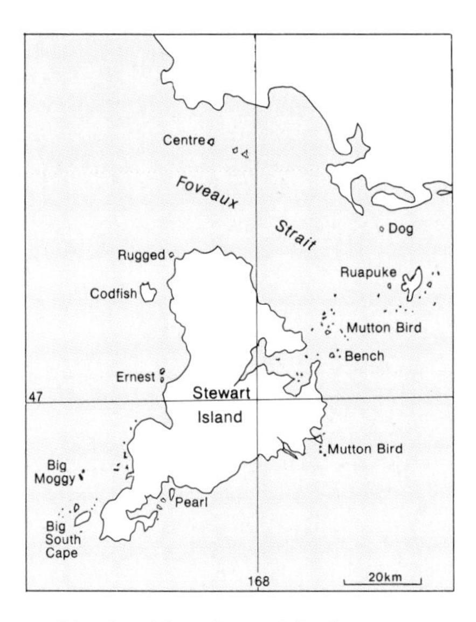

Above: *Stewart Island and its adjacent islands.*

Right: *The Fiordland crested penguin is mainly found along the shores of Foveaux Strait and the west coast of the South Island.*

beak is orange and its plumage blue-black. It spends early winter at sea, coming ashore in July and leaving again in February. Moulting takes place whilst ashore, some distance away from its breeding area. While this is happening it remains well concealed in caves or bush in an attempt to reduce the nuisance of sandflies, which are as much a worry to birds as they are to humans.

Small islands in Breaksea Sound are the only known habitats of a species of glossy black skink. *Leiolopisma acrinasum* is diurnal, living among stones, driftwood and seaweed on these remote islands. They grow to an overall length of 20 cm, have strong legs and a slender head with pointed snout. The back and sides are flecked with small spots of iridescent blue or green, while the underside is a soft greeny-grey.

The sooty shearwater is the commonest of the five species of New Zealand shearwaters, and in season, the most numerous seabird in southern New Zealand waters. In calm weather it flutters and glides but is stiff-winged in rough.

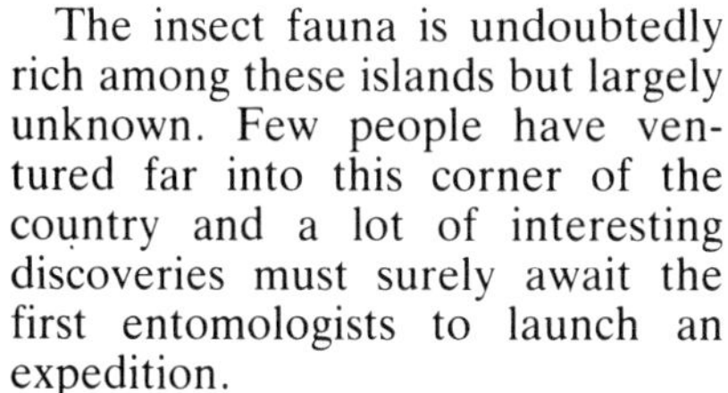

The insect fauna is undoubtedly rich among these islands but largely unknown. Few people have ventured far into this corner of the country and a lot of interesting discoveries must surely await the first entomologists to launch an expedition.

Of the five shearwaters common to the coasts of New Zealand, the sooty shearwater, or muttonbird, is by far the most numerous, especially in our southern waters. Small breeding colonies range from the Three Kings Islands and Cavalli Islands, but the greatest density is on the outlying islands of Stewart Island and the islands further south.

The muttonbird never normally comes to land except to breed. It winters in the Northern Hemisphere and breeds in the Southern. The nightly homecoming of the muttonbird has often been described as one of the marvels of the animal world. Countless numbers arrive from just before nightfall when they can be seen whirling and twisting in the sky. They circle noiselessly over the headland containing their burrows; a thud is heard, followed by a soft rustle – the first muttonbird has landed; more land, then the thuds increase until it becomes continuous. Once down, they scuttle to their burrows, and after an hour or so the arrivals gradually decrease until by midnight only the occasional straggler is to be seen in the air. The departure is more concentrated as they line up, just before dawn at the various 'take-off' points – places open enough and with sufficient height for them to become airborne.

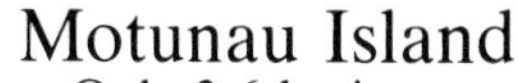

Motunau Island

Only 3.6 ha in area and the only island off the Canterbury coast, yet it is the home of some 20,000 breeding fairy prions and smaller

populations of several other seabirds. Rabbits were eradicated between 1957 and 1962 and, in the absence of grazing stock, the vegetation is recovering well. Tussock now provides suitable nesting sites for sooty shearwaters, white-faced storm petrels, and white-flippered penguins.

Permission to land is required from the Commissioner of Crown Lands, Christchurch.

Green Island

This tiny volcanic cone off the coast 10 km south of Dunedin, is rather insignificant except for an amazing variety of seabirds. Approaching the island one is quite likely to see an albatross, mollymawk or giant petrel. On landing, the large colony of Stewart Island shags and gulls take to the air in raucous disarray. Yellow-eyed penguins peer suspiciously at dusk and lumbering fur seals pretend not to notice.

The area is soon to be designated a Native Reserve.

Stewart Island

Really the smallest of New Zealand's main islands (but demanding individual treatment because of the number of animal and plant species peculiar to it). Stewart Island covers an area of 172,200 ha. The coastline is indented with many bays and beaches and several quite considerable harbours. Most of the island is forested, and this may be attributed to the high rainfall, mild winters and good soils. Some of the world's most beautiful ferns, scented native orchids and temperate plants flourish in the island's mild climate.

At the beginning of the present century it was realised that the primeval plant covering and the accompanying bird life was worth protecting. Several reserves have been gazetted and now over 105,000 ha are set aside in this manner. Although the vegetation has, to varying degrees, been modified by deer and opossum, Stewart Island remains a unique forested wilderness.

Opossums were introduced for their fur in 1890, red deer in 1901 and 1902 and whitetail deer in 1905. These liberations had disastrous effects. Rata trees have suffered considerably through the opossums, and deer have caused much damage to ground cover, especially soft ferns.

The whitetail deer is a native of the Americas, where many subspecies range from Canada south to Peru. The New Zealand herds are the descendants of a species found in the United States, and originally released on Stewart Island and near Lake Wakatipu. A timid but often curious deer with rather delicate facial features, the whitetail is so named because of its habit, when alarmed, of holding its large tail erect like a bushy flag. The conspicuous tail is lined underneath with long white hair that may act as a guide for juveniles when the parent bounds through the dense undergrowth. On Stewart Island its range is generally restricted to the coastal fringes, where it may sometimes be seen on isolated beaches foraging for seaweed.

Both deer and opossums are now regarded as noxious, but dogs, cats and rats are also responsible for much of the harm done to the birdlife. The ship rat population is extremely difficult to control, feral cats play havoc with nesting birds, especially the smaller petrels on some offshore islands, and dogs and cats are the main reason for the dwindling numbers of weka.

The Stewart Island fernbird is found at sea level as well as on the high hill tops. Its preferred habitat is usually fern or low scrub where it can be seen running, mouse-like, rapidly in and out of petrel burrows looking for insects.

However, many birds, subspecies of which are exclusive to Stewart Island, and birds seldom seen on the mainland, still charm the visitor. The Stewart Island robin, still seen frequenting the forest margins south of Paterson Inlet and adjacent islands, is slightly smaller than the South Island robin, the upper surface darker and the undersurface a greyish white.

There are five recognised subspecies of the once widespread fernbird, and though its numbers have been greatly reduced by the draining of wetlands and the burning-off of scrub and grass, it is still plentiful where conditions remain suitable. The fernbird's flight is weak and on the high windswept hilltops of Stewart Island its presence is obvious. It prefers low manuka scrub or fern in or near swampy ground, where it builds its nest in a clump of rushes. The cup-shaped nest, made of grasses and lined with a few feathers, often of the pukeko, normally contains a clutch of three eggs which are a pinky-mauve with tiny dark spots.

Sharing this exposed habitat is the largest of the dotterels that breed in New Zealand. Whereas the New Zealand dotterel in the northern part of the North Island prefers hot sandy beaches and rolling dunes, on Stewart Island it is mainly confined to the hilltops above the bushline. Here they build their nests among low-growing subalpine vegetation and stunted, wind-beaten carpets of manuka, sometimes descending to tidal flats to feed. A few breed on the sandy sweep of Mason Bay where creek beds cut through the dunes and banded dotterels, pipits and paradise ducks nest in close proximity.

The Stewart Island brown kiwi, one of the largest kiwis, a rich rusty brown in colour, is often heard and sometimes seen at Mason Bay. Although the kiwi is quite likely to be seen on dull days, a close approach may be made at night with torch in hand, when this fearless bird

Above: *The antler formation of the whitetail buck is distinctive. Brow tines which occur with most deer antlers are absent on the whitetail.* Right: *A female Stewart Island brown kiwi about to enter the burrow which is excavated mainly by the male.*

can be observed as it plods at a steady pace along the top of the beach probing in the sand and leaving behind its distinctive footprints. When fossicking for food with its long sensitive bill it appears at times to lean on it as if half-asleep.

The discovery of kakapo on Stewart Island in early 1977 brought renewed hope for a unique species that has been in danger of extinction for some time. Kakapo feed on a variety of vegetable matter including the fruits of subalpine plants and the leaves and roots of a variety of herbs and grasses. The

blades of tussock grasses are chewed, and fibrous material is left hanging in loose balls or scattered on the ground in compact pellets. These become bleached by the sun, thus providing a noticeable indication of the birds' presence. The future of this large nocturnal ground parrot remains in doubt; the apparent decline of the Fiordland population in recent years, and the scarcity of females, gives no cause for optimism. Perhaps the Stewart Island population will save the species.

Several species of shags coexist on the stacks and islets around Stewart Island; the pied shag and the little shag are abundant; the blue shag is seen in Paterson Inlet,though today, most of its ancestral nesting sites on the rocky ledges of vertical cliffs are deserted. The Stewart Island shag is increasing in numbers. The big green-bronze birds and their pied mates nest in large, close groups such as Whero Island, and can be seen on the Otago coast including the peninsula. Little and pied shags nest in trees on Paterson Inlet while spotted shags prefer the vertical cliffs of the inlet, and rugged rocks on the west coast.

Yellow-eyed penguins nest in the shelter of coastal bush, the little blue penguin in rocky holes or even under houses where their 'gossiping' may be heard on summer nights.

The forest, (especially in the northern half of the island), abounds with bush birds. Tuis and bellbirds, parakeets and cuckoos, pigeons, kakas and the brown creeper are all present. Smaller birds include fantails, tomtits, grey warblers and a large representation of introduced finches. Oystercatchers and herons patrol the foreshore.

Both of New Zealand's only native land mammals – the bats – survive on Stewart Island. The long-tailed bat is seen as it flies noiselessly about on summer evenings, flitting in and out of hollow trees just after dusk. The slightly larger short-tailed bat, with its plump, mouse-like body and wider wing-span is restricted to some of the offshore islands, and may be in danger of extinction due to rats.

Stewart Island with its associated islets has been an important stepping-stone in the distribution of

The blue shag in breeding plumage. It is a subspecies of the spotted shag.

The yellow-eyed penguin is of a genus and species found only in New Zealand and may be a relict species of the ancient penguins. Its range extends from Stewart Island up the east coast of the South Island as far as Akaroa.

invertebrates between mainland New Zealand and the subantarctic islands. These volcanic remnants are set in a notoriously turbulent ocean hence the chances of an organism surviving are small, whether it is airborne or transported by driftwood. Having landed, it must live long enough to find a mate and a suitable habitat. Very few insects which occur on Campbell Island (650 km south) can be found through the Aucklands (500 km), the Snares (190 km), Stewart Island, (25 km) to the South Island. Others, such as the weevil *Oclandius*, which have been isolated since the Ice Ages, have evolved separate species on each of the islands but can be traced through their close relationships back to mainland New Zealand.

The shore-plover now seems to be confined to South-east Island (Rangatira) in the Chathams, although wind-blown stragglers may sometimes reach other islands.

Codfish Island

This island of 1214 ha off the north-west coast of Stewart Island is a scenic reserve with a wealth of birdlife. Few islands of its size could rival it for its variety of seabirds, a number of which breed on the island, and an equal number can be sighted in the vicinity. Mollymawks glide effortlessly offshore. Long burrows in steep banks on the island itself are the homes of mottled petrels, sooty shearwaters, Cook's petrels (of which Little Barrier Island is the only other known nesting ground) and broad-billed prions. The latter is reasonably common all around our coasts, breeding on the islands of Foveaux Strait, and is easily recognisable by its wide curved bill. Yellow-eyed and Fiordland crested penguins nest in low fern or scrub and among rocks at the edge of the bush. The bush itself supports numerous birds, not least the island's own subspecies of fernbird.

The host of small islands that make up Dusky Sound, one of the thirteen sounds of the Fiordland coast, where the stoat has become a threat to bird and animal life

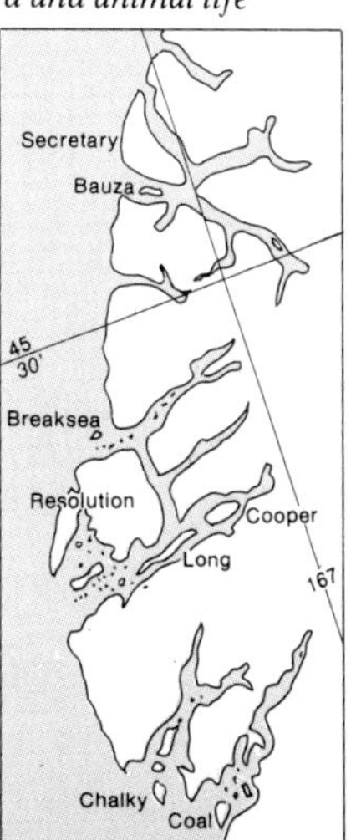

THE FUTURE

Since the year 1600 – when records were first kept – over 225 species and subspecies of birds and mammals have been removed from the face of the earth. Some became extinct because of natural factors such as climatic change, but the vast majority of more recent extinctions are the result of man's cruelty, greed and foolishness. According to one American environmental scientist, humans have, in the last 2000 years, exterminated about 3 per cent of the world's known mammal species – but over half have occurred since 1900!

The dominance of man has left less and less room for the planet's other inhabitants. Today something like 900 species of animal life are seriously threatened, and this figure includes a large number of New Zealand species. Of the 318 species or subspecies of rare or endangered birds listed in the IUCN (International Union for Conservation of Nature and Natural Resources) Red Data Book, approximately 11 per cent are from New Zealand and the outlying island groups (Kermadec, Auckland, Campbell, Antipodes, Bounty and Chatham). New Zealand has this unenviable record partly because of the special circumstances which make the fauna of isolated islands or archipelagos particularly vulnerable to the threat of extinction, and partly because of the treatment New Zealand has received at the hand of man, especially European man. (In discussing New Zealand fauna which has become extinct, or is today rare or endangered, the emphasis is on the birds, for in the pre-human environment the only other native animals were a few lizards, three species of frog and two species of bat. These have survived better proportionally than the avifauna).

Nearly all of the land birds in the world which have become extinct in the last 150 years have lived on islands. The ecological characteristics of long-isolated oceanic islands are such that they tend to have a depauperate land and freshwater avifauna, the species have broader ecological niches than their counterparts on large land masses and many are endemic. New Zealand has been widely separated from other lands for approximately 70 million years. Of the 77 (indigenous and endemic) land and freshwater bird species listed as breeding in the New Zealand region 60 per cent are endemic.

According to the New Zealand Checklist, 45 species were extinct at, or soon after, the advent of European man. The extinct forms are the moas, seven waterfowl (two of which, however, are extant in Australia), six rails, two hawks, two eagles, one snipe and one crow. The causes are still not fully understood. Some authorities believe that climatic and vegetational changes had a major impact whereas others believe that the early Polynesians were responsible.

It is not known exactly how many species became extinct between the arrival of Polynesian man approximately 1000 years ago and that of European man 200 years ago, but it seems that at least 34 of the 45 which were extinct by 1800 were alive during the early Polynesian era, because their bones have been found in middens and campsites. The extent to which the early Polynesians were implicated in these extinctions is difficult to ascertain. It is believed that they were often primarily responsible because they caused major ecological changes through the large scale burning of some of the original vegetation, by hunting, and by introducing the kiore and the Polynesian dog. Certainly the kiore is known to have caused a decline in tuatara and lizard populations, and on islands where there are kiore there is generally a lack of small petrels such as storm-petrels. However, there are difficulties in trying to ascribe the responsibility primarily to the Polynesians and their dogs and rats, because the rates of extinction in the North, South and Chatham Islands prior to 1800 were similar. This is despite the fact that the South Island had a very much smaller Polynesian population, mainly confined to the northern, eastern and southern coasts.

Whatever changes occurred in the New Zealand environment before 1800, the modifications in the 200 years following European settlement have been great and extremely rapid. During this period eight full species, seven subspecies and numerous populations have become accepted as extinct or presumed to be so. These figures do not fully demonstrate the extent of the changes because a further 23 species or subspecies have become endangered or rare including the Stewart Island snipe and Chatham Island bittern. The impact on the Chatham Islands' fauna has been particularly severe with five species or subspecies having populations of only 100 or so birds.

Endangered and Rare Species

There are then a number of causes which account for the perilous state of our fauna, and avifauna in particular, at the present time, namely:

Destruction of habitat

Large-scale forest clearing, the draining of swamps and burning-off to break in new land for farming are all introduced conditions which only a few of the indigenous species of birdlife can tolerate. New Zealand quail owes its extinction to just this reduction of habitat. Since 1840 approximately 66 per cent of the original forest cover has been removed.

Introduction of new species into New Zealand

The introduction of predacious mammals in particular has disturbed the former balance of avian life in all habitats – no tract of countryside is now without its stoats and rats which are the worst enemies of the smaller native birds. The black and brown rats are responsible for destroying nests; stoats for the decimation of ground-dwelling, flightless species and young birds; while opossums and deer attack the tree cover, de-barking and defoliating in their browsing and foraging. Wild dogs are responsible for the deaths of hundreds of wekas and kiwis, and feral cats, the extinction – amongst other crimes – of the Stephens Island wren. Introduced birds have to some extent displaced native birds at the urban/rural, rural/forest interfaces.

The collection of species for private collections and museums

Large numbers of huias, stitchbirds, piopios and other birds were – like the legendary auk – appreciably reduced in numbers by 19th century collectors. (The great auk was in fact literally collected to extinction. It once inhabited rocky

Defoliated trees on Rangitoto Island, damage caused by the ubiquitous opossum.

islands in the North Atlantic Ocean and was killed in great numbers by commercial fishermen and whalers landing on the islands. Then, as the species became rare, museum collectors and scientists rushed in to obtain specimens and eggs before the bird disappeared altogether. The last two auks known to exist were killed in 1844, for a collector).

Hunting

Hunters must be credited with causing a certain reduction in the numbers of some vulnerable species. One instance of this concerns the kakapo, formerly abundant in the Makarora Valley but cleared out there by workmen employed in forming the road through the area. The kakapo was also a victim of the early Maori and his dog, who regarded the meat as a delicacy.

Some elements of the New Zealand avifauna are more vulnerable to extinction than others. In European times the endemic fauna has suffered an extinction rate approximately eight times that of the remainder of the indigenous (native but not endemic) fauna. Similarly, of the species that are listed as endangered or rare, 19 per cent are endemic and only 4 per cent indigenous.

New Zealand is fortunate in having some offshore islands that have remained almost wholly unmodified by European settlement and introductions, and are still rodent-free. If it had not been for these islands New Zealand's list of extinct species could have been greater and would have included the stitchbird, North and South Island saddleback, and the shore-plover. It is possible that some petrels on offshore or outlying islands may have also in the past bred on the mainland. The kakapo is another species surviving better on an island.

The unmodified islands are valuable not only because they support examples of the primitive biota but because many of them can be used as refuges to which certain threatened species may be transferred so that free-ranging and self-sustaining populations can establish. Whilst this technique has been used successfully recently for the conservation of two island races of saddlebacks, and the black robin, the little spotted kiwi and the eastern weka were also saved from extinction by liberations on islands at the beginning of this century. The little spotted kiwi has been sighted only twice on the South Island mainland since 1940 and is now believed to be extinct there, and the eastern weka became extinct on the mainland about 1920.

Any list of endangered birds must to some extent be subjective. The following gallery of endangered wildlife divides the species or subspecies into five categories –

1. Probably already extinct;

2. Endangered – in considerable danger of extinction and special protective measures are necessary to ensure continued survival;

3. Rare – numbers low. In some instances the population appears to

The grey opossum. The opossum has caused considerable damage to forests throughout the country as well as to crops such as turnip, and competes with stock for pasture and clover.

be stable or increasing, in others the population trends are unknown and they may be declining;

4. Vulnerable – although not at present endangered or rare the forms in this category can be considered vulnerable because of a very restricted breeding distribution;

5. A fifth category can be allocated to those birds which are visitors to our shores. While not endangered or rare in their native habitat (usually Australia) they are sufficiently uncommon here to warrant mention.

In placing forms into the rare or endangered categories, due regard has been taken of the population size, population trend over the last decade, the distribution, and whether or not the population can survive with or without some form of direct management.

1. Probably Already Extinct

North Island Bush Wren
(Xenicus longipes stokesi)

This weak-flying forest bird is known from only two specimens collected last century. Unconfirmed sightings were reported from the Urewera National Park in 1955.

South Island Bush Wren
(Xenicus l. longipes)

An inhabitant of high altitude forests this subspecies was widely distributed in the mountain areas of the South Island from Nelson to Fiordland. The decline in numbers appeared to have occurred after 1880. It is now extremely rare. The last reports were from Milford Sound, Fiordland in 1965 and Nelson Lakes National Park in 1968.

Stead's Bush Wren
(Xenicus l. variabilis)

This race was fairly abundant on the islands off the southwest coast of Stewart Island (Big South Cape and Solomon). It is also believed to have inhabited the forests of Stewart Island until the ship rat reached the islands in fishing boats in 1962. Before the species disappeared in 1967, six were transferred to nearby rat-free Kaimohu Island. In 1967 and 1972 two were observed there but an inspection in 1977 failed to find any.

Laughing Owl
(Sceloglaux albifacies)

This species was once widely distributed in the eastern part of the South Island and the lower half of the North Island. It began to disappear in the North Island in the early 19th century or before (only one specimen was ever collected), and rapidly became scarce in the South Island after about 1880. No confirmed sightings have been made since 1914 but there have been unconfirmed sightings in Fiordland.

Stewart Island Snipe
(Coenocorypha iredalei)

Recorded sightings of this species on three islands west of Stewart Island (Big South Cape, Solomon and Pukaweka) were last made in 1964, but the bird is now regarded as extinct, having been eliminated by ship rats.

South Island Kokako
(Callaeas cinerea cinerea)

Once well distributed over much of the South Island forested areas particularly in the west, this orange wattled subspecies was declining at a rapid rate by the turn of this century. Despite special searches, reports of this subspecies remain unconfirmed, and the South Island kokako is generally regarded as extinct.

North Island Thrush
(Turnagra capensis tanagra)

Once widely distributed in the North Island this species suffered a rapid decline with the last official report being made in 1887. Occasional reports of sightings have been made in the Urewera National Park but field investigators have failed to confirm them.

South Island Thrush
(Turnagra c. capensis)

This subspecies was once exceedingly abundant in the forests of the South Island and was regarded as common on the west coast of the South Island during the gold digging era of the 1860s. By 1870 however, the species was considered to be fast disappearing. Recent unconfirmed sightings have been made in West Otago in 1963 and Fiordland in 1962. However, because both the North and South Island races are similar in appearance to the ubiquitous song thrush, confirmation is needed.

2. Endangered –

In considerable danger of extinction and special protective measures are necessary for continued survival.

Chatham Island Taiko
(Pterodroma magentae)

This medium-sized petrel was rediscovered in 1978 when two were captured and released and another one was seen on the same night. Formerly the species was known only from one specimen and numerous subfossil bones. It was once very numerous and a food source of the early Polynesian inhabitants of the Chathams. A small remnant breeding population apparently exists but has not yet been located. The decline of the taiko has probably come about primarily because of destruction or modification of its habitat and the introduction of cats, dogs, rats and pigs.

Kakapo
(Strigops habroptilus)

The kakapo was once widespread

in both the North and South Islands and Stewart Island. Prior to European settlement the species was in decline and it had become localised in the North Island and extinct in the east of the South Island. In the early part of this century it was plentiful in Fiordland, the West Coast, and parts of Nelson and Southland. It is now restricted to the vicinity of Milford Sound which is largely deer-free, and Stewart Island. After five years (1973-1978) of intensive searching of the Milford hinterland, only 11 kakapo had been found, all of which have been males. In 1977 the species was rediscovered on Stewart Island after having been last seen there in 1949. On the basis of sign, at least 30 kakapo are considered to inhabit an area of several thousand hectares of scrubland east of the Tin Range. The distribution on Stewart Island is still being investigated.

Orange-Fronted Parakeet

(Cyanoramphus malherbi)

Records show that this species has never been common; it has only been recorded in a few localities this century – all in the South Island (Fiordland, Nelson and Canterbury) and on Stewart Island. Because it often occurs in mixed flocks, particularly with the superficially similar yellow-crowned parakeet, there is difficulty in determining the true status of the species. It still persists in small numbers in the north Canterbury and southern Nelson montane areas.

Chatham Island Robin

(Petroica traversi)

Formerly distributed over most of the Chatham Islands, a population of about 20 managed to persist in a 4 ha forest remnant on Little Mangere Island. Forest destruction and cats are believed to have been the main causes of the extinction on other islands in the Chatham group. The forest habitat on Little Mangere has been deteriorating over the past decade and this culminated in the species declining to only seven birds in 1975. By 1977 all birds were transferred to the nearby Mangere Island. In 1977 the population rose to nine but it is back to seven again at present.

Takahe

(Notornis mantelli)

The present distribution is confined to approximately 650 sq km of bush and alpine tussock grassland of Fiordland National Park comprising the Murchison Mountains and a small area to the north. The shrinkage in range predated the arrival of Europeans. It was known from the North Island only from subfossil bones, although it is suspected that it may have persisted in low numbers in the lower half of the North Island in the Ruahine Ranges. Similarly in the

South Island, subfossil material indicates a much wider distribution than in recent times. In European times it has been positively identified only from Fiordland, but it is possible that it persisted for some time in the Nelson area. The present population is estimated to be 200-250. A recent decline in numbers has given concern for its continued survival. The main part of the bird's range, the Murchison Mountains, was gazetted a special area with restricted entry soon after the takahe was rediscovered in 1948.

Black Stilt

(Himantopus novaezealandiae)

Formerly widespread but more abundant in the South Island than in the North Island, it is now rare with breeding being restricted to the upper reaches of the Waitaki River in the South Island. The population suffers to some extent from hybridisation with the pied stilt. The population of pure black stilts probably does not exceed 70. Hydro-electric development and farming practices are severely modifying the breeding habitat but attempts are being made to provide alternative habitat to replace that lost through hydro-electric work, and to control or exclude predators from nesting areas. In the 1977-78 season only six juveniles were produced. The low productivity has resulted from high predation by ferrets and cats.

Little Spotted Kiwi

(Apteryx oweni)

Once occurring on both islands in high rainfall areas, this kiwi has, since the turn of the century, been restricted to the western half of the South Island, occurring from Marlborough Sounds to Fiordland. It was once the commonest kiwi on the West Coast of the South Island and formed the bulk of the thousands of kiwi skins that were shipped to Britain to be made into fashion muffs and feather collars between 1860 and 1890. Since 1940 there have been only two confirmed records and an intensive field research programme during the last two years has failed to locate any. These findings do not necessarily imply extinction on the mainland, but at present the species is at best sporadically distributed and in extremely low numbers. Fortunately, through the foresight of one of the early directors of the Forest Service, five birds were released on the island sanctuary of Kapiti at the beginning of this century and today their offspring are believed to number at least two hundred. In 1980 three birds were found on D'Urville Island and have been transferred to Maud Island where two survive. It is uncertain whether this is a subspecies or a separate species.

Chatham Island Pigeon

(Hemiphaga novaeseelandiae chathamensis)

This is confined to the small broadleafed forest remnants on the southern half of Chatham Island. Its reduction has apparently come about because of the destruction of forest, cat predation, and hunting by man. Currently the population is approximately 50, but it is reproducing adequately wherever the habitat is satisfactory, however the area of forest is declining because of its conversion to pasture.

Efforts are being made to reserve and manage some major forest areas on both Chatham and Pitt Island to which it is proposed to transfer some birds at a suitable time.

3. Rare –

Numbers low. In some cases the population appears to be stable or increasing, in others the population trends are unknown and numbers may be declining.

Forbes' Parakeet

(Cyanoramphus auriceps forbesi)

Formerly widespread on the Chatham Islands it is now confined to about 6 ha of bush on Little Mangere Island and to a similar sized remnant on nearby Mangere Island. Prior to the mid 1970s this parakeet occurred only on Little Mangere and here in 1938, the population was estimated at about 100 birds. Another estimate in the early 1970s put its numbers at about only 10 pairs. Since then it has spread to Mangere and the total population is now about 40 birds.

North Island Kokako

(Callaeas cinerea wilsoni)

Last century this subspecies was found in most forested districts of the North Island and Great Barrier Island. Since European settlement

the distribution and numbers have shrunk substantially. At present it occurs in isolated populations of varying size in some widely separated localities; the main stronghold being podocarp-hardwood forests in the central North Island (West Taupo, Rotorua). Small numbers occur on Great Barrier Island, Coromandel, Hunua Ranges, Raglan, Waitomo and Taranaki, East Cape, Urewera and possibly Northland.

Some populations in the central North Island are known to be declining and this is undoubtedly linked to the diminution and alteration of the forests. Some forested areas appear to be too small to support long-term self-sustaining populations and too isolated for contact and exchange with neighbouring populations.

Parts of the West Taupo forests which are a stronghold for kokako have been logged for timber until recent years but as a result of public pressure and a request from the New Zealand Wildlife Service a three year moratorium has been declared by the Government. This allows time for research to be carried out on the feeding and habitat requirements of the bird.

North Island Saddleback

(Philesturnus carunculatus rufusater)

During the last century this subspecies was plentiful and widely distributed throughout the North Island, on Little and Great Barrier Islands and Cuvier Island and Hen Island. Although it is believed to have existed in the East Cape area until 25 years ago, there have been no recent sightings and it is generally recognised that this bird now persists naturally only on Hen Island. Although the kiore also occurs on Hen Island, saddlebacks are common and maintaining their numbers.

Since 1964, saddlebacks from Hen Island have been released on nearby Whatapuke, Red Mercury, Cuvier, Fanal, Marotiri and Kawhihi Islands. Each liberation has resulted in the establishment of a viable population with the total now numbering approximately 1000.

South Island Saddleback

(Philesturnus c. carunculatus)

The pattern of diminution of this subspecies is very similar to that of its North Island counterpart. At the beginning of European settlement, it was widespread in forest areas – D'Urville Island, Stephens Island, and Stewart Island and outlying islands. By 1900 it was virtually extinct on the South Island mainland, but managed to persist on

three islands off the southwestern tip of Stewart Island (Big South Cape, Solomon and Pukaweka Islands). In 1962 the ship rat reached these islands but before the saddlebacks disappeared, some were transferred to two nearby islands. The progeny of some of these have been moved to five other nearby islands. Although the population is now only about 180, the survival of the subspecies appears to be assured so long as the islands can be kept rat-free.

Stitchbird
(Notiomystis cincta)

Formerly widely distributed in the North Island and on Great Barrier Island, the species now persists only as a small but apparently stable population on Little Barrier Island. It is believed that stitchbirds became extinct on the mainland about 1885. The species does not appear to be in immediate danger of extinction, particularly as feral cats on Little Barrier Island have been eliminated and with the decline in the number of rats, numbers have increased considerably. In 1980 the Wildlife Service successfully transferred 30 birds to Hen Island, and a further 16 in 1981, resulting in a current population as numerous as it was initially on Little Barrier Island.

Codfish Island Fernbird
(Bowdleria punctata wilsoni)

Restricted to a small area of swamp on the upper slopes of Codfish Island (population c.500) this race was formerly widespread on the island. It has probably been affected by predation by kiore, wekas and by opossums modifying their habitat.

Brown Teal
(Anas aucklandica chlorotis)

Once widespread throughout the North and South Islands, and common on some near offshore islands and Stewart Island, the range of this duck has been markedly reduced. Between 1915 and 1930 the bird disappeared from much of the mainland. At present the species persists locally in very small numbers in parts of Northland (max. 400), Hauraki Gulf islands (mainly Great Barrier Island), Coromandel Peninsula, and in Fiordland and Stewart Island. In Northland, Fiordland and Stewart Island the population is continuing to decline and in Fiordland, sightings over the past 20 years have usually been of only one or two birds or pairs without young. On Stewart Island no positive sightings have been recorded since 1969 whereas 10 years earlier they were numerous.

The greatest population (c.1000) is found on Great Barrier Island and this population is believed to be stable but under threat from increasing agricultural and recreational development. Two other subspecies of this duck occur on the outlying islands of Auckland and Campbell Island. The Auckland Island race *(A. a. aucklandica)* survives in satisfactory numbers on a number of islets but on the main Auckland Island itself it is absent because of feral cats and pigs.

The Campbell Island race *(A. aucklandica nesiotis)* is confined to one small offshore island and probably numbers less than 50 birds. On the main island it has obviously suffered from predation by Norway rats and cats; in fact this race was rediscovered on Dent Island only in 1975 after not being seen for approximately 30 years.

Blue Duck
(Hymenolaimus malacorhynchos)

An inhabitant of fast-flowing stable forest streams, this species was formerly widespread throughout New Zealand from sea level to high in the mountains. The clearing of the native forests has inevitably caused its local extinction. Erosion induced by overgrazing of introduced ungulates in river catchments has also seriously affected their habitat by increasing sediment loading. Although the duck is still widespread, there has been a steady but substantial decline. It is now confined to mountain areas with main strongholds being in the Urewera, Tongariro and Fiordland National Parks and the North West Nelson Forest Park.

King Shag
(Leucocarbo c. carunculatus)

This subspecies is confined to a small area of the Marlborough Sounds in the Cook Strait region. They have never been known to be numerous or widespread. In recent years the population has probably increased slightly as additional breeding colonies have established. The present estimated population of 200-300 is believed to be about the same as that discovered in 1773. During the 19th century a decline in population probably occurred after large numbers were collected for the feather trade and for museum specimens. In recent years there has been less indiscriminate shooting of shags in general. The nesting colonies are easily disturbed during the breeding season, the main danger being from pleasure craft and fishermen.

Chatham Island Petrel
(Pterodroma hypoleuca axillaris)

This little-known petrel is confined to South East Island where the small breeding population is of undetermined status. It is possible that breeding sites of this species are being taken over by the broad-billed prions.

Auckland Island Rail
(Rallus pectoralis muelleri)

Confined to Adams Island, where a few sightings have been made in recent years. The exact status of this subspecies is unknown at present but its numbers are unlikely to be more than 500. It is apparently not on the main island of the Aucklands because of the presence of cats.

Chatham Island Oystercatcher
(Haematopus chathamensis)

Still occurs throughout the Chatham Islands but the main concentration (13 pairs) is on South East Island (one of the two island reserves in the group). It has probably never been a numerous species but numbers declined particularly on the two main islands because of predation by cats and perhaps by the introduced eastern weka preying on the eggs and young. The total population by a recent reliable census is 25-30 pairs.

Shore Plover
(Thinornis novaeseelandiae)

Once widespread throughout the coastal areas of New Zealand and the Chatham Islands this species now persists only on South East Island. Although the population is about 120 (90 adults, 30 juveniles) it would be extremely vulnerable to predation. Attempts to establish it on Mangere Island have so far failed because many of the transferred birds returned to their natal island. As long as predators do not reach the island the species is secure, but as insurance it is essential that further attempts be made to establish it on other predator-free islands.

Auckland Island Dotterel
(Charadrius bicinctus exilis)

This newly described subspecies breeds on the upper slopes of the main island and Adams Island and winters mainly on Enderby Island. On the main island it is subject to predation by cats. The population is thought to be less than 200.

Chatham Island Snipe
(Coenocorypha pusilla)

Though formerly restricted to South East Island where it was extremely rare in the mid 1950s, with the reservation of the island and the removal of its sheep the population has increased dramatically. In the early 1970s about 20 were transferred to Mangere Island. These have established successfully and now can be found over the whole island with the population exceeding 200. Probably because of increased numbers some birds have colonized the Star Keys, a small island 5 km north of South East Island.

4. Vulnerable –

Although not at present endangered or rare, the forms in this category can be considered vulnerable because of very restricted breeding distribution.

Hutton's Shearwater
(Puffinus huttoni)

The population is believed to number more than 5000 and could be as great as 20,000. The breeding ground which is restricted to alpine grassland (above 1220 m) of the Seaward Kaikoura and possibly the Inland Kaikoura Mountains was discovered only in 1965. At sea the bird is found in the Cook Strait and along the east coast of the South Island. Migration also occurs to South and Western Australia. The main nesting grounds have been gazetted a Nature Reserve. The trampling of introduced browsing animals (goats, deer and chamois) has collapsed many burrows and caused erosion of the friable loamy areas where they breed. However these noxious animals are now being controlled by the New Zealand Forest Service. Evidence of predation on adult birds in the breeding grounds by stoats, rats and native falcons has been observed.

Westland Black Petrel
(Procellaria westlandica)

This species, like the Hutton's shearwater, has a restricted breeding range and is confined to the Barrytown area of Westland. Latest population trends indicate that the species is increasing slightly. It seems to benefit by feeding behind fishing trawlers in a similar way to the fulmar *(Fulmarus glacialis)* in the United Kingdom.

In addition, there are a number of birds which while not facing extinction, are at risk because of reduction of habitat. These include the New Zealand dotterel, the kaka, wrybill plover, black-fronted tern, fernbird, brown creeper, grey teal, Stewart Island snipe, black petrel, fairy tern, crested grebe, New Zealand bittern and Caspian tern. In most cases their breeding grounds are specialised and limited and man's intrusion into these areas threatens an existence which is being maintained only, and is not expanding.

5. Visitors To Our Shores

New Zealand must also count among its uncommon and low-in-numbers avifauna, a range of winged migrants from Australia and the South Pacific. Some, like the crested grebe have now settled here on inland lakes and ponds. Others, like the royal spoonbill, golden plover, turnstone, white ibis, red-necked stint and avocet remain itinerants.

The Endangered Small Animals and Invertebrates

Geckos and Skinks

There are 33 species of native lizard in New Zealand, in the gecko and skink families, and all are at risk from human development of the natural habitat and from the carnivorous animals man has introduced. Little is known of their habits or exact numbers, but unless given active protection and management, at least half a dozen species face continuing decline and extinction. These are the giant skink *(Leiolopisma grande grande)*, confined to an Otago range; the Great Barrier skink *(L. homalonotum)*; Falla's skink *(L. fallai)*, confined to the Three King's Islands; the rough-scaled gecko *(Heterophotis rudis)*, Nelson, Marlborough and North Canterbury; the Stephens Island gecko *(Hoplodactylus stephensi)*, Stephens Island only; *H. nebulosis*, Stewart Island only; and the Lewis Pass green gecko *(Heterophólis poecilochlorus)*, Lewis Pass – Reefton area.

The Lewis Pass green gecko is diurnal though seldom seen in the hottest hours of the day.

The Stephens Island gecko, found only on Stephens Island in Cook Strait.

Hamilton's frog, the rarest of the three native frogs, inhabits the damp vegetation of the forest floor, burrowing deeply in drier weather.

Hochstetter's frog, the commonest of the three species of native frogs, frequents stream banks and is found in many parts of the northern half of the North Island.

Native Frogs

Archey's Frog
Leiopelma archeyi

Hamilton's Frog
L. hamiltoni

Hochstetter's Frog
L. hochstetteri

All three native frogs are restricted in their habitat, Archey's frog to a limited area of the Coromandel Peninsula; Hamilton's frog to Stephens and Maud Islands, and Hochstetter's also to the Coromandel area and near Auckland. The greatest danger to the frogs is destruction of habitat: they rely heavily on their skins for gaseous exchange, and are very susceptible to drying out. Unlike every other frog species the New Zealand native frogs have no tadpole stage in water, instead all development takes place within the egg.

Native Land Snails

At least 45 species and isolated populations of the large endemic land snails are considered to be endangered or vulnerable through changes to habitat brought about by human settlement and disturbance by wild goats and pigs. Indeed several populations of the flax snail *(Placostylus hongii)* have recently become extinct. Their salvation lies, as with so much of New Zealand's endangered fauna, in transference of a sufficient number of individuals to suitable undisturbed, predator-free islands and in restoring and protecting their original habitats.

Short-Tailed Bat
(Mystacina tuberculata)

The lesser short-tailed bat is considered to have a reasonably-sized population distributed through forests in the North Island, on Little Barrier Island, and in the northern

Archey's frog is found only on the Coromandel Peninsula, sometimes near streams though usually under stones or logs away from standing water.

and western parts of the South Island. However, the greater short-tailed bat, once found on Stewart Island and two of its immediately outlying smaller islands, is now considered extinct, the cause being the same as that which threatens the lesser species – predation by rats and stoats and destruction of habitat through modern farming and forestry techniques.

Freshwater Fish

Endemic fish populations have been little studied until recently. Prior to this, at least one, the grayling has become extinct, and several others have approached endangered status:

The giant kokopu and the smaller species, the short-jawed kokopu are now rare in developed inland waters due to destruction of their specialised habitat and competition with introduced trout. The black mudfish and the Canterbury mudfish both have small populations and are restricted in range, the first to swamps of the upper North Island, the second to the Canterbury area.

The 1980 Wildlife Amendment Act grants official protection to certain invertebrates such as some species of wetas, flies, grasshoppers, beetles, bugs, spiders, harvestmen, and land snails, whose populations are at risk for a variety of reasons. Some may soon be the subject of conservation projects.

Protection of endangered species

In New Zealand all native birds with three exceptions, the black shag, harrier hawk and the southern black-backed gull are protected (either fully or partially) under the Wildlife Act as are a number of small native animals and invertebrates. The New Zealand Wildlife Service has the statutory responsibility of enforcing the protective regulations, of preserving and managing habitats and of planning and pursuing research programmes. Naturally, with restraints on manpower and finance a scale of priorities is needed to determine where the greatest effort on endangered species should be directed. This priority ranking is based on the taxonomy, numerical status, distribution, stability of the habitat and the population trend. The following examples illustrate the different procedures being adopted with saving New Zealand endangered species.

Island Transfers of Black Robin

As mentioned earlier, some unmodified offshore islands hold the key to the survival of many of New Zealand's endangered species. The Wildlife Service is in the process of evaluating all offshore islands to ensure that the most valuable ones are protected by reservation and, if required, a management programme will be instituted. In the past this has involved the removal of noxious animals and, in others, re-planting vegetation to speed up regeneration.

The most critical of conservation projects which the Wildlife Service has undertaken involved the translocation of the black robin. The black robin had disappeared from the main islands of the Chatham Island group by 1871 and, after the introduction of cats to Mangere Island in the 1890s it became confined to Little Mangere Island. For 70 years a population of approximately 20 birds survived on about 4 ha of low forest on the summit of this small and precipitous island; but during the last decade there has been a rapid deterioration in this habitat, caused partly by a prolonged dry spell lasting for several years which subjected the vegetation repeatedly to salt damage. Furthermore this has been exacerbated by damage to regeneration arising from the activities of great numbers of sooty shearwaters.

All this has caused a steady decrease in the robin population, and by September 1976 it was reduced to two pairs and three males. It was evident that unless the birds were moved from Little Mangere the species would soon become extinct, since the small population had a high adult mortality (33 per cent per annum) and a high reproductive failure.

In September 1976 the Wildlife Service transferred five (three males and two females) to a small but rapidly regenerating coastal forest in the south-eastern slopes of nearby Mangere Island. Several years prior to the release of the robins the Wildlife Service had removed the sheep and planted thousands of suitable native trees and shrubs. Financial assistance for the planting programme had been provided by the Royal Forest and Bird Protection Society of New Zealand. In the 1976-77 breeding season one chick was produced but one adult male died. In February 1977 the two remaining male robins were removed from Little Mangere Island. In 1977-78 two chicks were

Black robin fledgling on Little Mangere Island.

Kapiti Island, looking south. Thought to be part of an ancient land bridge connecting both main islands, now a haven for numerous native and exotic birds.

produced but one pair disappeared. The current population numbers nine birds but, more importantly, there are three pairs instead of the original two at the time of transfer.

Re-establishment of Brown Teal

The brown teal is the rarest species of New Zealand's native waterfowl. Although not immediately in danger of becoming extinct its distribution is restricted and slowly diminishing. The Wildlife Service is attempting to rectify this by reintroducing captive-reared birds back into areas where, earlier this century, it was common.

To determine whether captive-reared stock would survive in the wild, nine teal bred at the Mt Bruce Native Bird Reserve and one from the wild were released onto Kapiti Island in 1968. Within four months of release one female was observed with a brood of four and in subsequent years other females were seen with young, elsewhere. In addition teal almost certainly originating from Kapiti Island, were seen at two localities on the mainland – one adjacent to Kapiti, the other about 80 km away.

The programme for breeding brown teal in captivity began in 1973 with only a small number of birds held at Mt Bruce and by a few private aviculturalists. From this small stock, 20-25 birds were raised and liberated annually. More recently, more private aviculturalists have joined the scheme and a small number of pairs were removed from Great Barrier Island in order to provide them with breeding stock. At present 35 pairs are held in captivity. Since the inception of the programme 143 teal have been liberated, 123 of them on three highly fertile sand dune lakes north of Wellington. It was on these lakes that the birds had bred as recently as 1930. Now some of the liberated birds have dispersed and established themselves on other small lakes nearby. And some have bred successfully.

The future of this venture depends on the careful choice of liberation sites and the ability to release numerous birds annually. In parts of their current wild range, brown teal are showing an ability to adapt, albeit slowly, to feeding on pasture and breeding on farm stock-ponds. But it is the heavily vegetated margins of water bodies that remain its prime requirement and this emphasises the niche which the animal primarily occupies. It is believed that the liberation of captive-reared birds may not be sufficient to permanently re-establish the species at various sites; so supplementation with wild birds from Great Barrier Island may be needed to stabilise the population. This may require the annual removal of about 50 birds. Research is in progress to establish whether this number can be removed without detrimental effect.

The brown teal is the rarest of New Zealand's native ducks.

Conservation

In Nature, all forms of life are dependent on one another. Man is only one among millions of species and must cease to consider the rest of creation as freely available for his own purposes. Being alone in his capacity to affect the biosphere consciously, it is essential that he should set himself guidelines in so doing.

There is a dreadful permanence about the extinction of a species. The extermination or drastic reduction in numbers of any form of native wildlife is a tragic circumstance for the country concerned. It is, however, doubly tragic when such species contribute to the distinctive character of a country and are, in many cases, unique and and irreplaceable.

The native wildlife of New Zealand falls into just such a category. Because of the long isolation of these islands and the absence of any mammalian predators a special evolutionary pattern was able to develop which produced, out of approximately 250 bird species native to New Zealand, many varieties found nowhere else.

The New Zealand Wildlife Act, 1953, and its subsequent amendments, legally provides absolute protection for every species of native bird not listed under its Schedules: *Wildlife Not Protected, Wildlife Partially Protected,* and *Wildlife Declared To Be Game.* This means that every native animal and almost every native bird is accorded complete protection, as is any new species arriving or establishing itself in the country (the welcome swallow from Australia is the most recent example).

The reservation of suitable habitat is considered an essential step in any programme of wildlife preservation, and many areas in New Zealand are set aside for this and allied purposes. These include National Parks and a considerable number of smaller areas set aside as Wildlife Sanctuaries, Flora and Fauna Reserves and Scenic Reserves. The legal provisions attaching to sanctuaries and reserves are stringent, and entry is prohibited without an authorising permit which, in the case of sanctuaries, is generally restricted to approved scientific study only. Many of New Zealand's offshore islands have some conservation status and are regarded as being of particular importance because, owing to their relative inaccessibility, rare species and distinctive communities which have not survived on the mainland continue to persist. Stephens Island (situated in Cook Strait), the main stronghold of the tuatara and one of the two known habitats of a rare species of native frog, is perhaps the best example.

The satisfactory management of wildlife habitat is the next step after reserving it, and essential specialised knowledge in this field is gained through research. All of New Zealand's major offshore islands have been surveyed and, as a result, proposals have been made for the creation of reserves where necessary, and for the application of various management programmes. These programmes include the removal of predators, the planting of soil-stabilising and sheltering native flora, the erection of protective fencing and, where deemed suitable, the reintroduction of native bird species. Transfer work is also carried out, as for example the South Island saddleback being saved from extinction a few years ago. In the mainland forests, a constant surveillance is being maintained in areas known to support native bird populations and which are likely to be endangered by development. In addition to this work, field trips are made to determine the presence and distribution of such rare species as the takahe and kakapo and also to check on reported sightings of other rare species.

For some very severely threatened birds adequate protection and management of habitat is not always possible in the time available. The establishment in captivity, therefore, of populations of native birds which have been reduced to small, localised numbers in the wild, not only ensures against their ultimate disappearance, but also allows for study which in other circumstances would be impossible. With these aims in view, the Mt Bruce Native Bird Reserve was established by the Government in 1961. It has since achieved considerable success in the twin fields of encouraging the breeding of rare native birds such as the North Island saddleback, both varieties of Antipodes Island parakeet, the blue duck and the eastern or buff weka, and research into their behaviour and ecology to isolate the causes of their decline and open the way for possible re-establishment in suitable areas of their former range. Takahe and kakapo have both been intensively studied at Mt Bruce, the former over a long period, and it is to the work carried out at this Reserve that we are indebted for most of our present knowledge of both species.

The Stewart Island brown kiwi, like the North Island subspecies, has a distinctly reddish brown plumage; the South Island subspecies is a duller brown.

Legislations; habitat acquisition; management programmes; research; these are essential tools in the cause of practical conservation, but will ultimately prove ineffectual if the cause they serve lacks public support. To make absolutely sure that it does not, Government Departments, Acclimatisation Societies, Conservation, and Ecology Groups as well as private individuals are all engaged in the ever-growing and indispensable work of education.

The Role of National Parks and Public Reserves in Nature Conservation

Together, national parks and public reserves make a major contribution to nature conservation in New Zealand and to the protection of representative examples of ecosystems. Between them, the 10 National Parks and more than 1000 reserves primarily devoted to nature conservation occupy a little over one-twelfth of the total land area of the country.

Polynesian and Old World and New World influences have all played a part in the evolution of today's system of national parks and public reserves. Polynesian influence was central to the establishment of New Zealand's first national park – Tongariro – with the gift to the nation in 1887 of the park's nucleus by the paramount chief of the Ngati Tuwharetoa people. Here, the Polynesian reverence for land was given statutory recognition through a New World concept drawn from the United States Government's establishment 15 years earlier of Yellowstone as the world's first national park.

Tongariro was to be the forerunner of more gifts where other Maori groups gave land as scenic reserves, at Lakes Rotoiti and Okataina, for example, and more recently the gift of Egmont National Park.

A system of reservation of natural areas complementary to national parks evolved with the Department of Lands and Survey given responsibility under the Land Act 1892 to reserve Crown land for the growth and preservation of timber, for the preservation of native fauna, for recreation and gardens, for thermal springs or 'natural curiosities or scenery of a character to be of national interest'. This applied only to Crown land, and authority to buy privately owned land for similar purposes was added by the Scenery Preservation Act 1903.

Through the years Parliament continued to enact legislation to establish and manage protected areas for nature conservation and public recreation, notably drawing the threads together in the Public Reserves, Domains, and National Parks Act 1928 and, in the 1950s, enacting separate legislation in the National Parks Act 1952 and the Public Reserves Act 1953. The current Reserves Act 1977 provides not only for management of all public reserves but for the purchase of private land for reserves.

Together, the National Parks Act and the Reserves Act (supported by the reservation provision of the Land Act 1948) form the basis for today's system of protected areas built up by successive governments setting aside public land, buying private land, and accepting gifts of land for the purpose. Inherent in it is public ownership of the protected areas, which have the status of either national parks or public reserves.

The National Parks Act defines national parks as 'areas of New Zealand that contain scenery of such distinctive quality or natural features so beautiful or unique that their preservation is in the national interest'. There are important elements of public use, for the Act states its purpose as preserving national parks in perpetuity 'for the benefit and enjoyment of the public'.

Northern royal albatross, guarding its chick at Twin Rock, Taiaroa Head.

Mt Bruce Native Bird Reserve in the Wairarapa.

The Act lays down a number of criteria for the management of national parks:

- They shall be preserved as far as possible in a natural state.
- Native flora and fauna shall as far as possible be preserved, and introduced flora and fauna shall as far as possible be exterminated unless otherwise determined.
- Sites of archaeological and historical interest shall as far as possible be preserved.
- The public shall have freedom of entry and access to the parks 'so that they may receive in full measure the inspiration, enjoyment, recreation, and other benefits that may be derived from mountains, forests, sounds, lakes, and rivers'. (This provision is subject to conditions necessary for preservation of native flora and fauna and for the welfare in general of the parks'.)

The Act provides for the establishment within national parks of special areas to give maximum protection to nature and wilderness areas to cater for those who seek isolation. In addition, the Act lists a range of permissible developments, including camping grounds, huts, hostels, accommodation houses, ski tows, roads, and tracks always subject to the overall aim of the Act.

Protected areas under Reserves Act 1977

The Reserves Act 1977 is important in complementing the National Parks Act. It provides for the Minister of Lands to classify reserves 'according to their principal or primary purposes' to ensure their 'control, management, development, use, maintenance, and preservation' for their appropriate purposes.

The seven classifications listed in the Act are:

Recreation reserves
Historic reserves
Nature reserves
Scientific reserves
Scenic reserves
Government purpose reserves
Local purpose reserves

In the context of nature conservation the most significant classifications are scenic, nature, and scientific reserves, but this does not mean that there are not strong elements of nature conservation in other classifications. Indeed, one of the features introduced in the Reserves Act in 1977 was the concept of reserves management for multiple use. For example, though

The kaka, earlier present in great numbers on the mainland, is now restricted to native forest areas of both Islands, and larger coastal islands, and appears reluctant to inhabit exotic plantations or settled districts.

local purpose reserves are 'for the purpose of providing and retaining areas' for 'educational, community, social, or other local purposes', the Act (section 23) says that 'where scenic, historic, archaeological, biological, or natural features are present on the reserve, those features shall be managed and protected to the extent compatible with the principal or primary purpose of the reserve'. Subject to the same proviso, the value of a local purpose reserve for soil, water, and forest conservation is to be maintained.

Similar provisions apply to other classifications of reserves but one group of reserves in the Government purpose classification deserves particular mention in the context of protected areas. Section 22 (2) says that Government purpose reserves may include reserves for wildlife management or other specified wildlife purpose.

Recreation reserves (section 17) are not only for the purpose of 'providing areas for the recreation and enjoyment of the public' but are 'for the protection of the natural environment and beauty of the countryside, with emphasis on the retention of open spaces and on outdoor recreational activities, including recreational tracks in the countryside'. As in historic reserves (section 18), scenic, archaeological, biological, geological, or other scientific features or indigenous flora or fauna or wildlife present on recreation reserves 'shall be managed and protected to the extent compatible with the principal or primary purpose of the reserve'.

Nature reserves

Nature reserves (section 20) are 'for the purpose of protecting and preserving in perpetuity indigenous flora or fauna or natural features that are of such rarity, scientific interest or importance, or so unique that their protection and preservation are in the public interest'. This clearly places emphasis on habitat for endangered or rare species and on areas where control on public access is necessary.

Entry to nature reserves is by permit, and the Act says that they must be so managed as to preserve them as far as possible in a natural state with 'indigenous flora and fauna, ecological associations, and natural environment' as far as possible preserved and 'exotic flora and fauna as far as possible . . . exterminated', except where the Minister otherwise determines.

Nature reserves equate to special areas in the national park context, and there are 71 of them totalling 179 101 ha. Many outlying and offshore island reserves fall into the nature reserve classification – the subantarctic islands (The Snares, Campbell, Auckland, Bounty, and Antipodes), the subtropical Kermadecs, and such offshore islands as The Three Kings, Poor Knights, Hen and Chickens, Little Barrier, Kapiti, and the Chetwodes. Mainland areas qualifying as nature reserves are Farewell Spit, the white heron colony area on the Waitangiroto, and the reserve at Castle

Awahou is an impressive scenic reserve on the Stratford-Ohura highway.

Rainbow Falls on the Kerikeri River, Northland, lie within the Rainbow Falls Scenic Reserve.

Hill, Canterbury, for the preservation of the rare buttercup, *Ranunculus paucifolius.*

Scientific reserves

The scientific reserve classification is closely allied to the nature reserve classification, but there is greater emphasis on research values and less emphasis on rare and endangered species.

Fifteen areas totalling 2052 ha have been classified as scientific reserves.

Scenic reserves

The numerically and physically largest category of protected areas under the Reserves Act falls into the scenic reserve classification, with 1008 reserves totalling 322 114 ha.

Scenic reserves provide for the public to have freedom of entry and access subject to conditions 'necessary for the protection and well-being of the reserve and for the protection and control of the public' and, except where the Minister otherwise determines, 'the flora and fauna, ecological associations, and natural environment and beauty shall as far as possible be preserved'.

The purposes of scenic reserves fall into two categories: reserves to be managed essentially as natural areas, and reserves capable of enhancement by the introduction of indigenous or exotic flora.

These reserves include a great variety of protected areas – gullies of indigenous forest in suburbia; large tracts of forest flanking the Wanganui River, in the Marlborough Sounds, the Buller Gorge, the Lewis Pass, and on Stewart Island; and geological features such as the Waitomo, Nile River, and other cave systems and the pancake rocks and limestone country near Punakaiki.

National reserves

Where, in the Minister's opinion, a reserve protects values of 'national or international significance' (section 13) that reserve may be declared a national reserve, which gives a direct link with the term 'national park'.

An act of parliament is needed to change the classification of such a reserve. A further link with national park status is in the provision authorising the Minister to set apart the whole or any part of a reserve as a wilderness area (section 47), with similar provisions for management as for wilderness areas in national parks.

The compatibility of national parks and national reserves is apparent, and the classifications of scenic, scientific, and nature reserves as well as historic, recreation, and Government purpose (wildlife management) reserves, further emphasise the wide-ranging scope of protected areas provided for under the National Parks Act and the Reserves Act.

Sanctuaries

In order to conserve our native fauna, it must be protected against predation and destruction of habitat. Rats, stoats and deer must be kept in check and useful habitats maintained in as near natural conditions as possible.

Some especially endangered species may even be able to be transferred from inhibiting surroundings to other, more encouraging habitats, or populations may be maintained and bred in captivity to improve the species' chances of survival. (While the ultimate objective with any endangered species is to maintain populations in the wild, the business of improving habitats and the selection and

Fur seals by the surf at Waikowhai Bluff on the West Coast.

preparation of new habitats can be protracted, and keeping a population in captivity may ensure its survival until its existence in the wild is no longer in jeopardy).

National Parks and scenic reserves can play an effective general role in conserving fauna, but it has been proved necessary to create special sanctuaries and reserves for the particular care of threatened bird and animal life.

The Wildlife Service is responsible for many of the sanctuaries and refuges around New Zealand, amongst which are: Duffers Reef, White Rocks and The Brothers and Trio Islands in the Marlborough Sounds; Cosgrove and Rainbow Islands, off the eastern Southland Coast; Sentinel Rock and Stephens Island in Cook Strait; Mayor and Karewa Islands, Bay of Plenty; and Motunau Island off northern Canterbury. Here may be found such protected colonies as tuatara (as on Stephens Island), and the king shag (Duffer's Reef).

Then there are the flora and fauna reserves such as Little Barrier and the Chetwode Islands (under the jurisdiction of the Hauraki Gulf Maritime Park Board and Marlborough Sounds Maritime Park Boards respectively), and special reserves such as the Mt Bruce Native Bird Reserve and Taiaroa Head (Otago) albatross colony. Perhaps the best known is the Mt Bruce Native Bird Reserve: located near Masterton.

The Reserve has three main functions –

1. To hold and breed in captivity endangered species of native wildlife.

2. To be a centre for study of native wildlife, particularly endangered species.

3. To provide for public education in wildlife conservation.

The Buller Gorge Scenic Reserve protects native forests flanking the gorge.

The Trounson Kauri Park Scenic Reserve provides walking tracks in a Northland kauri forest.

Before transferring birds to Mt Bruce all available information on their living requirements is studied. This provides some guidelines but the fact that many of our native birds have seldom, if ever, been kept in captivity before means that, for a start, a degree of experiment is required in their management. Whenever possible some experience is gained with a closely related but more numerous species before rare birds are captured. Once a decision has been made to take birds into captivity careful co-ordination is required to ensure that they undergo as little stress as possible during capture and transport.

The most urgent need once birds are at Mt Bruce is to see that they resume feeding. At first special dietary requirements are added to drinking water and a wide variety of natural and artificial foods is offered including, if possible, known preferences of the species in the wild. It is impossible, however, to duplicate exactly the birds' natural diet and, as a feeding pattern becomes apparent, a simplified diet of readily available foods is evolved.

The following species have been bred in captivity at Mt Bruce: takahe, weka, pukeko, North Island kiwi, little spotted kiwi, yellow-crowned parakeet, red-crowned parakeet, Antipodes Island parakeet, Antipodes Island red-crowned parakeet, Forbes' parakeet, tui, North Island saddleback, blue duck, brown teal, New Zealand scaup, grey teal, grey duck, paradise shelduck, shoveler.

The Tin Range area of the Pegasus Nature Reserve on Stewart Island is the habitat for the rare kakapo.

The Work of the Wildlife Service

The New Zealand Wildlife Service is a division of the Department of Internal Affairs. Though the Service as such is 30 years old, the Department (or its predecessor, the Colonial Secretary's Office) of which it is now a part has been responsible for the administration of wildlife matters for well over a century. In fact, the first item of legislation specifically dealing with New Zealand wildlife, was the Protection of Certain Animals Act – which made no mention of native species whatsoever – promulgated in 1861.

In 1913 the Department of Internal Affairs took over the management of freshwater fisheries in the Rotorua district, and in 1926 did the same in Taupo. In 1930 it became, in effect, the Acclimatisation Society in what is at present the Central North Island Wildlife Conservancy, and in 1945 took over the same functions in the present Southern Lakes Acclimatisation District. Official recognition that New Zealand had a noxious mammal problem (apart from rabbits) led to the establishment in 1931 of a Deer Control Section. The Department exercised the function of hunting and trying to control a wide range of mammals – but mainly deer – for a quarter of a century, until this responsibility was transferred to the Forest Service in 1956. The work done by the Government deer-cullers during those 25 years, was often under appalling and primitive conditions in the wildest and most remote parts of New Zealand.

In 1945 the Deer Control Section became part of the newly-founded Wildlife Branch: two years later its first scientist was appointed. From then on, the activities of the Branch became steadily more concerned with scientific research and with the conservation of native species, though the killing of introduced browsing and grazing mammals still accounted for the major activities of its staff. Added impetus to research and conservation resulted from the rediscovery of the supposedly extinct takahe in 1948; the New Zealand-United States Fiordland Expedition of 1949; and the controversy over government policy on noxious mammals in the early 1950s.

When the Deer Control Section was transferred to the Forest Service in 1956, the loss of the greater part of the Wildlife Branch's staff caused a material and psychological setback that lasted for about 10 years. However, the decade starting from about 1965 has seen a steady increase in members, responsibilities, resources, experience and reputation to the extent that the Wildlife Service (as it is now called) is well known overseas – as well as in New Zealand – for the quality of its work and has been called on by overseas governments to supply manpower and advice on a number of projects.

The legal basis upon which the work of the Service is founded is the Wildlife Act of 1953. As its preamble states, the Wildlife Act is 'an Act to consolidate and amend the law relating to the protection and control of wild animals and birds, the regulation of game shooting seasons, and the constitution and powers of acclimatisation societies'. Although there is a legal definition of 'wildlife' in the Act, a more accurate, (though not completely inclusive), and up to date description as far as the Service's responsibilities are concerned would be 'all free-living back-boned animals except marine fish, marine mammals and those mammals included in the Wild Animals Act of 1977 (i.e. deer, chamois, goats, thar, opossums, wallabies, pigs) as well as any domestic animals gone wild'.

Because of its responsibility for the management of the freshwater fisheries in the Central North Island and Southern Lakes Acclimatisation Districts, there is a second Act of great importance to the work of the Wildlife Service – the Fisheries Act (1980) Part II (Freshwater Fisheries); and insofar as it administers this Act, the Service works in close collaboration with the Fisheries Research and Management Divisions of the Ministry of Agriculture and Fisheries.

Another close relationship is that between the Wildlife Service and the various acclimatisation societies, those private sportsmen's organisations each responsible within its own district for administering (under delegation from the Ministers of Internal Affairs and Agriculture & Fisheries) the Wildlife Act and Part II of the Fisheries Act.

The headquarters of the Service is in Wellington, and there are main district offices in Rotorua, Queenstown, Auckland, Christchurch and Dunedin. In addition, wildlife officers are widely scattered about New Zealand – from Kerikeri to Invercargill. The Mt Bruce Native Bird Reserve, is an avicultural and research station for some of New Zealand's rarest fauna, and there are other research centres at Pukepuke Lagoon (for water fowl), and Kaikoura (for bush birds and predators).

The many functions of the Service make it convenient for it to be divided into nine sections whose main responsibilities are, briefly, as follows:

Administration This section supervises and co-ordinates the day-to-day running and planning of the whole organisation and ensures its co-ordination with all other relevant organisations.

Limestone formations at Castle Hill, North Canterbury, part of the reserve which protects the rare Ranunculus paucifolius.

New Zealand's 'living fossil', the tuatara, whose only relatives are as long gone as the age of the dinosaurs, was once widespread but is now restricted to islands off the north-east coast of the North Island, and Cook Strait.

Research The Service is basically scientific orientated and relies upon ecological research for guidance in almost all of its management policies. The scientists and technicians of the Research section, in close collaboration with the various kinds of field staff, supply the necessary ecological guide-lines.

Environment The Service has found it necessary to set up a section to specialise in assessing environmental impact statements, town and country planning legislation, and various other matters concerned with land and water use insofar as these things affect wildlife and freshwater fisheries.

Protected Fauna This section is responsible for the conservation and management of all protected wildlife. This entails study and management, including the translocation of endangered species. However, its work is not restricted to rare species but encompasses the management of habitat for all species. Its activities extend from time to time from the Kermadecs in the subtropics to Campbell Island in the subantarctic, and from Fiordland in the west to the Chathams and Antipodes Islands in the east.

In some of its so-called 'basic' and 'applied' research programmes, the Wildlife Service is guided by the Fauna Protection Advisory Council, a body of experts in various aspects of conservation, especially of wildlife, appointed by the Minister of Internal Affairs. The Service fosters collaboration with similar research organisations such as those in other government departments and in universities.

The service is also responsible, wholly or in part, for the setting up or administration of various kinds of reserves such as wildlife sanctuaries and other special reserves (Mokohinau Islands, Karewa Island, Maud Island, Taiaroa Head albatross colony, wildlife refuges, wildlife management reserves and

Wildlife officers rounding up a flock of paradise shelduck for banding at Benmore Station, Otago.

so on); and it carefully regulates the keeping of certain species of protected and partially-protected wildlife in captivity. By overseas standards New Zealand is virtually free of illegal trafficking in such species.

Fauna Survey Unit The Fauna Survey Unit is primarily involved with surveying areas of native forest which may be threatened due to clear felling or selective logging. The surveys determine the value of the habitat, and assess wildlife populations dependent on native forest. Recommendations are made, which endeavour to protect areas of habitat of sufficient size and diversity to ensure the survival locally of all common and rare bush bird species.

Game Management This section assesses changing resources of game birds (waterfowl and upland game) and their habitats. Working in close collaboration with the acclimatisation societies, it makes recommendations for the annual game seasons and for the acquisition and improvement of game bird habitats throughout the country.

Fisheries Management The responsibilities of this section are similar in essence to those of the Game Management section, but are restricted to the Central North Island and Southern Lakes Acclimatisation Districts which are the two most important trout-fishing districts in New Zealand. Not only are wild sport-fish populations managed, but the Service also maintains important trout hatcheries at Rotorua (Ngongotaha), Turangi and Wanaka.

Information Public education in wildlife matters is one of the two main functions of this section. It pursues this objective by making films, by collaboration in other ways with the broadcasting media, by supplying articles for newspapers and periodicals, by producing general Wildlife Service publications and arranging talks for schools and for other occasions. The section's other main function is to build up a national wildlife sound library and wildlife photographic collection.

General Duties All 'unspecialised' wildlife officers may be regarded as belonging to this section. They are responsible for law enforcement, public relations, data collection, and collaboration with various societies and organisations (public and private). In general, they are jacks-of-all-trades as far as wildlife and freshwater fisheries management and conservation are concerned. They are 'field' officers in the true sense of the word.

The general aims of the Wildlife Service are those laid down in the preamble to the Wildlife Act. The Service sees as its particular tasks:

1. The study of species that are rare or in danger of becoming extinct. Here the hope is that research will lead to methods of management that will increase the species range and numbers, so that their long-term survival is assured – even if, as a last resort, this should only be in captivity. Studies of this

Taiaroa Head on the Otago Peninsula, showing a spotted shag nesting area on the sea cliffs.

many of which are essential to the survival of several endangered species of plants and animals.

The Nature Conservation Council singles out wetlands as the areas most vulnerable to damage in New Zealand today. Swamps, salt-marshes, estuaries and the fringes of reeds and raupo around shallow lakes may not seem exciting to many people, but a wealth of wildlife depends on them, including several of New Zealand's less common birds, such as bitterns, egrets, crakes, and grebes, and the exotic visitors, the glossy ibis and the Nankeen night heron. Drainage and reclamation of land for farming or industrial purposes is diminishing our wetlands at an alarming rate. Damage by stock to the edges of streams and drainage canals, to salt

kind should also lead to a better understanding of the general causes of rarity and extinction. (Examples of such studies are those of the takahe, kakapo, saddleback and Chatham Islands robin.)

2. The study of ecological systems so that the numbers and habitats of benign, and often common, species may be conserved or increased for the enjoyment of all. (Examples: the forest fauna survey, the national wetlands survey, the offshore islands surveys.)

3. The study of species exploited by man. This is to ensure that harvest may be so geared to production that perpetual yield may be attained. (Examples: the paradise shelduck, black swan and mutton bird investigations.)

4. The study of pest species so that their numbers may be controlled (or even locally exterminated, if necessary) and their damage minimised. (Examples of this kind of project are the studies of bird hazards on airfields and the predations of certain introduced carnivores on some species of native birds.)

5. The study – for its own sake or for aesthetic or even 'sentimental' reasons – of species and their ecosystems that are unique to this country or that have particular scientific value. (Examples: those of kiwis, tuatara and native frogs.)

Animals and birds need space to live and there is little point in passing protective legislation for them if their environment is not also protected. New Zealand has a number of very different habitats,

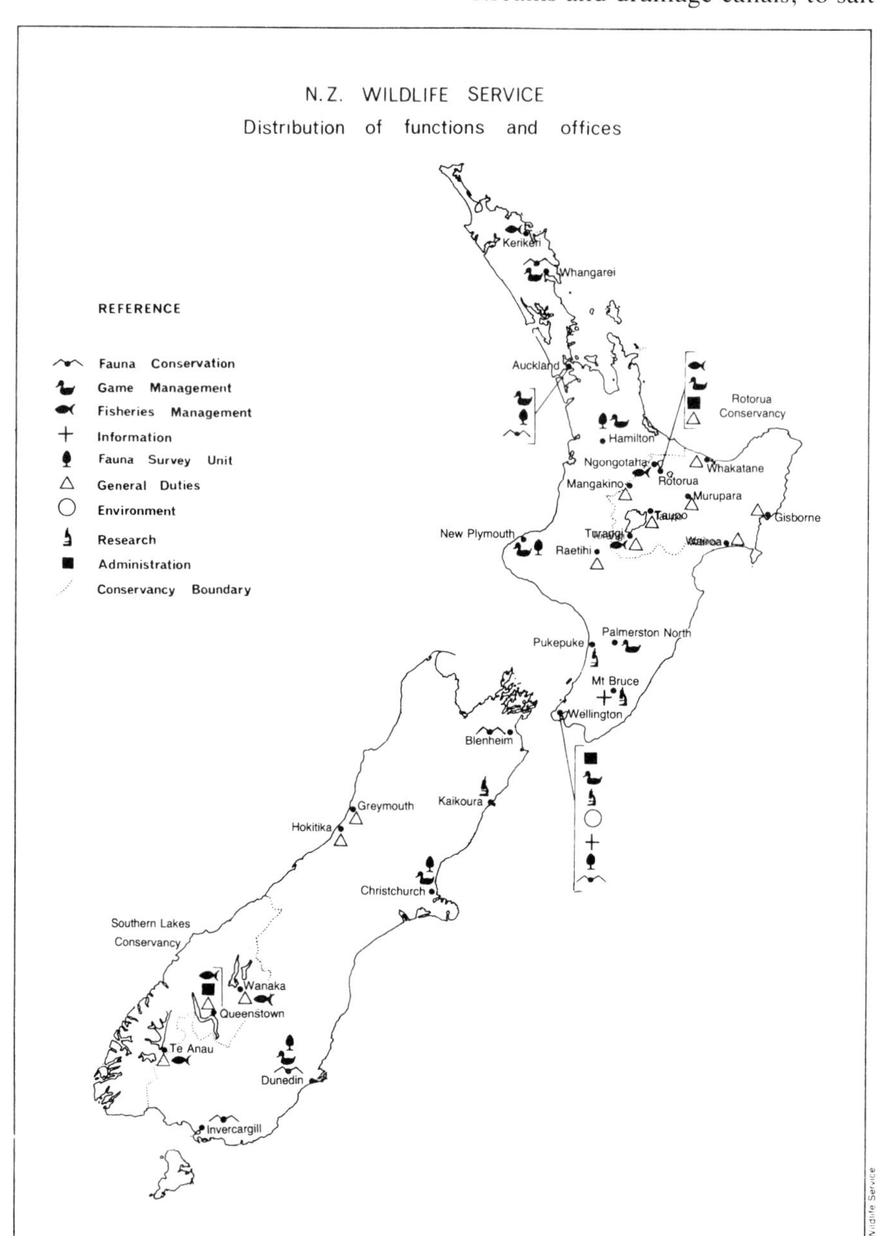

marshes and lake margins, seriously restricts the breeding areas of whitebait and other small fish.

The importance of estuaries in the overall economy of our coastal waters has only been realised in the last few years. Recent research has shown that the production of estuaries is equal to that of tropical forests and four times that of good rye-grass pastures. Although little of this vast productivity is of immediate use to us, it is nevertheless vital for our present and future prosperity. The often tiny and apparently insignificant plants and animals are all part of the long and complex food chain that supports commercial fisheries. At the base of the food chain are the submerged grasses, and the rushes and sedges from the salt marsh. Mangroves, sea-lettuce, large algae and the mud-dwelling micro-algae are also important producers. Direct grazing by marine animals consumes only about ten per cent of this material. Most of it dies and rots, and the resulting detritus is colonised by bacteria, fungi, protozoa and other micro-organisms. These in their turn become the basic food for crabs, shellfish, worms and fish.

Some fish can feed directly from the detritus and others eat the detritus-eaters. A 20 cm mullet will filter 1500 g of estuarine sediments a day, or more than 450 kg a year, to get its food.

These enormous numbers of shellfish, crabs and worms support huge flocks of birds, particularly gulls, ducks and many species of waders. The 4000-odd oyster-catchers studied at the Avon-Heathcote Estuary were found to have consumed in total an average of nearly one and a half million cockles a day during the winter months.

Estuaries are the main feeding and breeding ground for many commercially valuable species of fish. Natural beds of pipis, cockles, and mussels are exploited in many estuaries and commercial oyster and mussel farming is becoming increasingly important.

Estuarine sediments contain a higher percentage of very absorbent clays, and the filter feeding animals themselves remove enormous quantities of suspended matter from the water. This they deposit in the mud with their excreta. And their bodies add their accumulated store when they die. Because of this, estuaries are far richer in organic matter and nutrients than either the surrounding land or the open sea. But they are also natural traps for pollutants. Petroleum by-products, persistent pesticides, heavy metals and other toxic man-made chemicals are becoming increasingly concentrated in our estuarine waters and mudflats, with devastating effect on the wildlife.

Tussocks of curled snowgrass, snowberry shrubs, astelias, the great mountain daisy and pale midrib snowgrass cover this old moraine of the Hooker Glacier in the Mt Cook National Park. Snowgrass cover forms an excellent protective mantle, preventing erosion of steepland soils.

The plants and animals that live in the estuaries are extremely sensitive to changes in their environment. Unsatisfactory conditions can destroy a whole species. And the destruction of one seemingly insignificant species somewhere in the food chain can affect in turn the edible shellfish, commercial species of fish, and the seabirds and waders that feed on them.

Native grasslands are another habitat in danger of destruction. Huge areas have been lost because of oversowing with imported pasture grasses, topdressing and overgrazing. Tussock grassland can stand the occasional use of fire but not when it is followed by heavy overstocking. In the drive for pastoral production, the less obvious values of tussock are often overlooked. For example, it provides better protection against erosion than imported grasses. Again tussock grassland has the ability to intercept fog, or to trap water from the humid atmosphere and cause it to condense onto the ground. This property makes it very useful for increasing the water yield from a catchment.

Scrub areas, cut-over forest and other 'wasteland' comprises a third, under-rated but valuable habitat for many native birds. Brown creepers, tits, riflemen, fantails, grey warblers and bellbirds are some of the species that flourish in scrubland and forest margins.

A Northland marsh, one of the many wetlands so essential in maintaining the ecological balance.

Endangered species like the kokako are under pressure precisely because their habitat has shrunk almost to the minimum necessary to support a breeding population. The kokako demands a rather specialised environment – mixed tall trees, smaller trees and open areas – in which to find its food.

More thought can be given to modifying the landscape where it adversely affects a wildlife habitat, and to increasing available habitats of the more common fauna, birds in particular by planting suburban gardens with kowhais, eucalypts, New Zealand flax and many other nectar-bearing plants and trees which entice tuis and other nectar-loving birds. Goldfinches, chaffinches and white-eyes will flock to well-planted gardens at certain times of the year, and a few tall trees among the shrubs will encourage blackbirds and thrushes to settle in for the nesting season.

In the northern parts of New Zealand the aggressive Indian myna competes with its close cousin the starling. This does not please farmers and gardeners, who value the starling for its useful appetite but regard the myna (incorrectly) as a noisy but useless interloper. The starling can be encouraged by providing it with nesting boxes with an entrance just too small for the slightly larger myna to use.

Rapidly spreading in New Zealand since its arrival from Australia in the 1950s is the welcome swallow. It feeds on small flying insects, often over open water. Flocks of them can be seen perched on wires near streams and rivers. Also increasingly common in suburban gardens is the kingfisher, which will perch on a wire or a fence post with its tail flicking backwards and forwards with the regularity of a metronome. Despite its name, the kingfisher eats worms, grassgrubs, mice, lizards, insects and, only occasionally, small fish.

Conservation is not generally given priority over land development and utilisation. No national conservation policy exists, nor any single responsible authority. Such responsibility is currently divided among several government departments and other official organizations such as the Nature Conservation Council, Environmental Council and Commission for the Environment. The lack of a conservation policy hampers conservation measures.

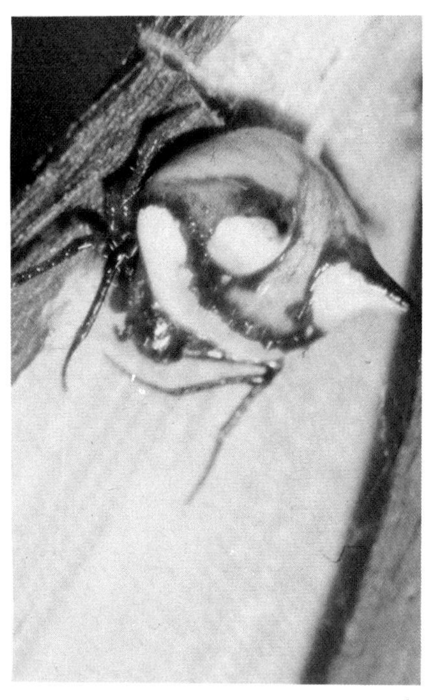

The colourful two-spined spider, an Australian immigrant, is now well-established in Auckland.

The introduced song thrush has settled successfully in all types of country and is one of the most common birds in New Zealand

SCIENTIFIC AND COMMON NAMES

PHYLUM COELENTERATA — Jellyfish, Sea anemones and Corals.

Aurelia aurita	Common jellyfish	
Isactinia tenebrosa	Red sea anemone	

PHYLUM PLATYHELMINTHES — Flatworms

Planaria spp	Flatworms	

PHYLUM MOLLUSCA — Molluscs

Family Athoracophoridae	Veined slug	
Amphibola crenata	Mud snail	Titiko
Cellana radians	Radiate limpet	Ngakihi
Chlamys zelandiae	Fan shell	
Chione stutchburyi	N.Z. cockle	Tuangi
Cominella adspersa	Speckled whelk	Kawari
Crassostrea glomerata	Auckland rock oyster	Tio para
Dosinia anus	Ringed dosinia	Tuangi-haruru
Haliotis australia	Southern or Silvery paua	Hihiwa
H. iris	Common paua	Paua
H. virginea	Virgin paua	
Helix aspersa	Garden snail	
Hyridella menziesi	Freshwater mussel	Kaeo
Latia neritoides	Freshwater limpet	
Melanopsis trifasciata	Decapitated snail	
Melarhapha oliveri	Periwinkle	Tatarepo
Mytilus edulis	Common mussel	Kuku
Nerita melanotragus	Black nerita	
Nucula hartvigiana	Nut shell	
Paphies australis	Pipi	Pipi
P. subtriangulata	Tuatua	Tuatua
P. ventricosa	Toheroa	Toheroa
Pecten novaezelandiae	Scallop	Tipa
Placostylus spp	Flax snails	
Potamopyrgus antipodarum	Dark water snail	
Sypharochiton pelliserpentis	Snakeskin chiton	
Turbo smargadus	Cat's eye snail	Ataata
Venericardia purpurata	Purple cockle	
Zeacumantus lutulentus	Horn shell	

PHYLUM ANNELIDA — Segmented worms

Abarenicola spp	Lugworms	
Spenceriella gigantea	Giant earthworm	

PHYLUM ECHINODERMATA — Echinoderms

Asterodiscus truncatus	Firebrick star	
Echinocardium austrole	Heart urchin	Kina
Evechinus chloroticus	Sea egg	Kina ariki
Patiriella regularis	Cushion star	Papatangaroa

PHYLUM ARTHROPODA — Arthropods
CLASS: CRUSTACEA — Crustaceans

Callianassa filholi	Ghost shrimp	
Eliminus modestus	Common barnacle	
Helice crassa	Mud crab	
Hemigrapsus edwardsi	Common rock crab	Papaka
Heterozius rotundifrons	Big-handed crab	
Leptograpsus variegatus	Large shore crab	Papaka
Mitella spinosa	Stalked barnacle	
Ovalipes punctatus	Swimming crab	
Pagurus novaezelandiae	Hermit crab	
Palaemon affinis	Common shrimp	Potipoti
Paranephrops planifrons	Freshwater crayfish	Koura wai
Patatya curvirostris	Freshwater shrimp	
Plagusia capensis	Red rock crab	Papaka
Talorchestia quoyana	Common sandhopper	

PHYLUM ARTHROPODA
CLASS: ARACHNIDA — Arachnids

Order Opiliones	Harvestmen	
Family Araneidae	Orbweb spiders	
Family Crenizidae	Ground trap-door spiders	
Family Dipluridae	Tunnelweb spiders	
Family Migidae	Tree trap-door spiders	
Family Salticidae	Jumping spiders	
Arachnura feredayi	Tailed forest spider	
Dolomedes Minor	N.Z. nurseryweb spider	
Latrodectus katipo	Katipo	Katipo
Poecilopachys australasia	Two-spined orbweb spider	

PHYLUM ARTHROPODA
CLASS: ONYCHOPHORA — Velvet worms

Peripatoides novaezealandiae	Peripatus	

PHYLUM ARTHROPODA
CLASS: DIPLOPODA — Millipedes

	Millipedes	

PHYLUM ARTHROPODA
CLASS: CHILOPODA — Centipedes

Cormocephalus rubriceps	Giant centipede	Hura

PHYLUM ARTHROPODA
CLASS: COLLEMBOLA — Springtails

Holacanthella paucispinosa	Springtail	
Hypogastrurus spp	Aquatic springtails	

PHYLUM ARTHROPODA
CLASS: INSECTA — Insects

Adalia bipunctata	Two-spotted ladybird	
Aenetus virescens	Puriri moth	Pepetuna
Agrius convolvuli	Convolvulus hawk moth	Hihue
Anagotus fairburni	Flax weevil	
Anisolabis laittorea	Seashore earwig	Mata
Anisops spp	Backswimmers	
Antheraea eucalypti	Gum emperor	
Apion ulicis	Gorse weevil	
Apis mellifera	Honey bee	
Arachnocampa luminosa	Glow worm	Piritana
Argosarchus horridus	Stick insect	Ro
Argyrophenga antipodum	Tussock butterfly	Mokarakare
Austrosimulium australense	Sandfly (N.Z. blackfly)	Namu
A. ungulatum	Sandfly (West Coast blackfly)	
Bassaris gonerilla	Red admiral	Kahukura
B. itea	Yellow admiral	
Bombus spp	Bumble bees	
Caedicia simplex	Green katydid	Kikipounamu
Chaerodes spp	Sand darkling beetles	
Chironomus zealandicus	Harlequin fly or midge	
Coelopa littoralis	Kelp fly	
Costachorema spp	Caddis flies	
Costelytra zealandica	Brown beetle (grass-grub)	Tutaeruru
Culicidae spp	Mosquitoes	
Cynthia kershawi	Painted lady	
Danaus plexippus	Monarch butterfly	Kahuku
Declana atronivea	Zebra moth	
Deinacrida fallai	Poor Knights weta	
Dorcus helmsi	Helms' stag beetle	
Dytiscidae spp	Diving beetles	
Ephemeroptera spp	Mayflies	
Erebida butleri	Butler's mountain ringlet	
Gymnoplectron giganteum	Giant cave weta	Weta
G. spp	Cave wetas	
Helleia salustius	Common copper	
Hemideina thoracica	Bush weta	Weta
Kikihia spp	Clock cicadas Tasman cicadas	Kihikihi
Lasiorrhynchus barbicornis	Giraffe weevil	Tuwhaipapa
Locusta migratoria	Migratory locust	Kapakapa
Lycaena boldenarum	Boulder copper	Mokarakare
Maoricicada cassiope	Cassiope cicada	Kihikihi
M. clamitans	Yodelling cicada	Kihikihi
M. oromelaena	Alpine black cicada	Kihikihi
Mecodema costellum	Carabid beetle	
Megateptoperla grandis	Stonefly	
Metacrias erichrysa	Mountain tiger moth	Pepe
Neocicindela spp	Tiger beetles	Kui

Nezara viridula	Green shield bug	
Notoreas brephos	Orange underwing moth	Pepe
Opifex fuscus	Saltpool mosquito	
Orcus chalybeus	Steel-blue ladybird	
Orthodera ministralis	Praying mantis	Ro whe
Percnodaimon pluto	Black mountain ringlet	
Pericoptus spp	Sand scarabs	Mumutaua
Pharmacus montanus	Mt Cook flea (weta)	
Phaulacridium marginale	Grasshopper	Kowhitiwhiti
Philanisus plebeius	Marine caddis	
Pieris rapae	Cabbage white	
Saldidae spp	Pondskaters	
Selidosema productata	Celery pine moth	Pepe
Sigara arguta	Water boatman	
Sigaus australis	Mountain grasshopper	
Teleogryllus commodus	Black field cricket	Kihikihi kai
Triamescaptor aotea	Mole cricket	
Uropetala carovei	Giant dragonfly	Kapokapowai
Wiseana spp	Porina moths	Porina
Xanthocnemis zelandica	Red damselfly	Kihitara
Xanthorrhoe orophyla	Barred upland looper moth	Pepe
Zelandoperia spp	Stoneflies	
Zizina otis labradus	Common blue	

PHYLUM CHORDATA
CLASS: AGNATHA — Chordates, Jawless fish

Geotria australis	Lamprey	

PHYLUM CHORDATA
CLASS: TELEOSTEI — Bony fish

Anguilla australis	Short-finned eel	Hao
A. dieffenbachii	Long-finned eel	Kuwharuwharu
Carassius auratus	Goldfish	
Cheimarrichthys fosteri	Torrent fish	Papamoko
Ctenopharyngodon idella	Chinese grass carp	
Galaxias argenteus	Giant kokopu	Kokopu
G. brevipinnis	Koaro	Koaro
G. divergens	Dwarf galaxias	Taiwharu
G. fasciatus	Banded kokopu	Para
G. gracilis	Dwarf inanga	Hiwi
G. maculatus	Inanga	Inanga
G. paucispondylus	Alpine galaxias	
G. postvectis	Short-jawed kokopu	Kokopu
G. prognathus	Long-jawed galaxias	
G. vulgaris	Common galaxias	Kokopu
Gambusia affinis	Mosquito fish	
Gobiomorphus basalis	Cran's bully	Pako
G. breviceps	Upland bully	Pako
G. cotidianus	Common bully	Pako
G. gobioides	Giant bully	Pako
G. hubbsi	Blue-gilled bully	Pako
G. huttoni	Red-finned bully	Pako
Neochanna apoda	Brown mudfish	Hauhau
N. burrowsius	Canterbury mudfish	Hauhau
N. diversus	Black mudfish	Hauhau
Oncorhynchus nerka	Sockeye salmon	
O. tshawytscha	Quinnat salmon	
Perca fluviatilis	Perch	
Poecilia latipinna	Sailfin molly	
Retropinna retropinna	Common smelt	Ngaiore
Rhombosolea plebeia	Flounder	Patiki
Salmo gairdnerii	Rainbow trout	
S. salar	Atlantic salmon	
S. trutta	Brown trout	
Salvelinus fontinalis	Brook char	
S. namaycush	Mackinaw	
Scardinius erythrophthalmus	Rudd	
Stokellia anisodon	Stokell's smelt	
Tinca tinca	Tench	

PHYLUM CHORDATA
CLASS: AMPHIBIA — Amphibians

Leiopelma archeyi	Archey's frog	
L. hamiltoni	Hamilton's frog	
L. hochstetteri	Hochstetter's frog	
Litoria aurea	Green tree frog	
L. ewingi	Whistling frog	

PHYLUM CHORDATA
CLASS: REPTILIA — Reptiles

Heteropholis gemmeus	Jewelled gecko	
H. manukanus	Marlborough green gecko	
H. poecilochlorus	Lewis Pass green gecko	
H. rudis	Rough-scaled gecko	
H. stellatus	Starred gecko	
Hoplodactylus duvauceli	Duvaucel's gecko	
H. granulatus	Forest gecko	Moko-papa
H. maculatus	Common gecko	Moko-papa
H. nebulosus		Moko-papa
H. pacificus	Pacific gecko	Moko-papa
H. stephensi	Stephens Island gecko	Moko-papa
Leiolopisma acrinasum	Fiordland skink	
L. aeneum	Copper skink	
L. fallai	Falla's skink	
L. grande grande	Giant skink	
L. grande otagense	Otago skink	
L. homalonotum	Great Barrier skink	
L. infrapunctatum	Spotted skink	
L. lineoocellatum	Speckled green skink	
L. macgregori	MacGregor's skink	
L. moco		
L. oliveri	Oliver's skink	
L. ornatum	Ornate or Forest skink	Moko-moko
L. pachysomaticum	Thick-bodied skink	
L. smithi	Smith's skink	Moko-moko
L. suteri	Black shore skink	
Naultinus elegans elegans	Auckland green tree gecko	Kakariki
N. elegans punctatus	Wellington green gecko	
N. grayi	Northland green gecko	
Sphenodon punctatus	Tuatara	Tuatara

PHYLUM CHORDATA
CLASS: AVES — Birds

Acanthisitta chloris	Rifleman	Titi-pounamu
Acridotheres tristis	Myna	
Alauda arvensis	Skylark	
Alectoris chukar	Chukor	
Anarhynchus frontalis	Wrybill	Ngutu parore
Anas chlorotis	Brown duck	Pateke
A. gibberifrons	Grey teal	Tete
A. platyrhynchos	Mallard	
A. rhynchotis	Shoveler	Kuruwhengi
A. superciliosa	Grey duck	Parera
Anthoruis melanura	Bellbird	Korimako
Anthus novaeseelandiae	N.Z. pipit	Pihoihoi
Apteryx haasti	Great spotted kiwi	Kiwi
A. australis	Brown kiwi	Kiwi
A. oweni	Little spotted kiwi	Kiwi
Ardea novaehollandiae	White-faced heron	Matuku-moana
Arenaria interpres	Turnstone	
Athene noctua	Little owl	
Aythya novaeseelandiae	N.Z. scaup	Papango
Botaurus poiciloptilus	Bittern	Matuku
Bowdleria punctata	Fernbird	Matata
Branta canadensis	Canada goose	
Calidris canutus	Knot	Huahou
C. ruficollis	Red-necked stint	
Callaeas cinerea	Kokako	Kokako
Carduelis carduelis	Goldfinch	
C. flammea	Redpoll	
Chalcites lucidus	Shining cuckoo	Pipiwharauroa
Charadrius bicinctus	Banded dotterel	Tuturiwhatu
C. dominicus	Golden plover	
C. melanops	Black-fronted dotterel	
C. obscurus	N.Z. dotterel	Tuturiwhatu
Chlidonias hybrida	Black-fronted tern	Tara
Chloris chloris	Greenfinch	
Circus approximans	Harrier hawk	Kahu
Coenocorypha aucklandica	N.Z. snipe	Tutukiwi
Corvus frugilegus	Rook	
Cyanoramphus auriceps	Yellow-crowned parakeet	Kakariki
C. malherbi	Orange-fronted parakeet	
C. novaezelandiae	Red-crowned parakeet	Kakariki
Cygnus atratus	Black swan	
C. olor	Mute swan	
Dacela novaeguineae	Kookaburra	
Egretta alba	White heron	Kotuku
E. sacra	Reef heron	Matuku-moana
Emberiza cirlus	Cirl bunting	
E. citrinella	Yellowhammer	
Eudynamis taitensis	Long-tailed cuckoo	Koekoea
Eudyptes pachyrhynchus	Fiordland crested penguin	
Eudyptula albosignata	White-flippered penguin	
E. minor	Blue penguin	Korora
Falco novaeseelandiae	N.Z. flacon	Karewarewa
Finschia novaeseelandiae	Brown creeper	Pipipi
Fringilla coelebs	Chaffinch	
Gallirallus australis	Weka	Weka
Gerygone igata	Grey warbler	Riroriro
Gymnorhina hypoleuca	White-backed magpie	
G. tibicen	Black-backed magpie	
Haematopus ostralegus	Pied oystercatcher	Torea

H. unicolor	Variable oystercatcher	Torea-pango
Halcyon sancta	Kingfisher	Kotare
Hemiphaga novaeseelandiae	N.Z. pigeon	Kereru
Himantopus leucocephalus	Pied stilt	Poaka
H. novaezealandiae	Black stilt	Kaki
Hirundo neoxena	Swallow	
Hydroprogne caspia	Caspian tern	Taranui
Hymenolaimus malacorhynchos	Blue duck	Whio
Larus bulleri	Black-billed gull	Tarapunga
L. dominicanus	Black-backed gull	Karoro
L. novaehollandiae	Red-billed gull	Tarapunga
Limosa lapponica	Eastern bar-tailed godwit	Kuaka
Lobibyx novaehollandiae	Spur-winged plover	
Lophortyx californicus	California quail	
Megadyptes antipodes	Yellow-eyed penguin	Hoiho
Meleagris galloparo	Turkey	
Mohoua albicilla	Whitehead	Popokatea
M. ochrocephala	Yellowhead	Mohua
Nestor meridionalis	Kaka	Kaka
N. notabilis	Kea	Kea
Ninox novaeseelandiae	Morepork	Ruru
Notiomystis cincta	Stitchbird	Hihi
Notornis mantelli	Takahe	Takahe
Numida meleagris	Guinea fowl	
Pachyptila vittata	Broad-billed prion	Parara
P. turtur	Fairy prion	Titiwainui
Passer domesticus	House sparrow	
Pelagodroma marina	White-faced storm petrel	
Pelecanoides urinatrix	Diving petrel	
Perdix perdix	Grey partridge	
Petroica australis	Robin	Toutouwai
P. macrocephala	Tomtit	Miromiro
Phalacrocorax carbo	Black shag	Kawau
P. carunculatus	King shag	
P. melanoleucos	Little shag	Kawaupaka
P. sulcirostris	Little black shag	Kawau
P. varius	Pied shag	Karuhiruhi
Phasianus colchicus	Pheasant	
Philesturnus carunculatus	Saddleback	Tieke
Platalea regia	Royal spoon bill	Kotuku ngutu-papa
Platycercus eximius	Rosella	
Podiceps rufopectus	Dabchick	Weweia
P. australis	Crested grebe	Puteketeke
Porphyrio melanotus	Pukeko	Pukeko
Porzana pusilla	Marsh crake	Koitareke
P. tabuensis	Spotless crake	Puweto
Procellaria parkinsoni	Black petrel	
Prosthemadera noveaseelandiae	Tui	Tui
Prunella modularis	Hedge sparrow	
Pterodroma cooki	Cook's petrel	Titi
P. macroptera	Grey-faced petrel	Oi
Puffinus bulleri	Buller's shearwater	
P. carneipes	Flesh-footed shearwater	
P. gavia	Fluttering shearwater	Pakaha
P. gavia huttoni	Hutton's shearwater	
P. griseus	Sooty shearwater	Titi
Rallus philippensis	Banded rail	Mohopereru
Rhipidura fuliginosa	Fantail	Piwakawaka
Sterna nereis	Fairy tern	Tara
S. striata	White-fronted tern	Tara
Stictocarbo punctatus	Spotted shag	Parekareka
Strigops habroptilus	Kakapo	Kakapo
Sturnus vulgaris	Starling	
Sula bassana serrator	Gannet	
Synoicus australis	Australian brown quail	
Tadorna variegata	Paradise shelduck	Putangitangi
Thinornis novaeseelandiae	Shore plover	
Turdus merula	Blackbird	
T. philomelos	Song thrush	
Turnagra capensis	N.Z. thrush	Piopio
Xenicus gilviventris	Rock wren	
X. longipes	Bush wren	Matuhi
Zosterops lateralis	Silvereye/White-eye	Tauhou

PHYLUM CHORDATA
CLASS: MAMMALIA Mammals

Alces alces	Moose	
Arctocephalus forsteri	N.Z. fur seal	Kekeno
Capra hircus	Goat	
Cervus canadensis	Wapiti	
C. elaphus	Red deer	
C. nippon	Sika	
C. timoriensis	Rusa	
C. unicolor	Sambar	
Chalinolobus tuberculatus	Long-tailed bat	Pekapeka
Dama dama	Fallow deer	
Erinaceus europaeus	Hedgehog	
Felis catus	Feral cat	
Hemitragus jemlahicus	Thar	
Hydrurga leptonyx	Leopard seal	Pakaka
Lepus europaeus	Hare	
Macropus bicolor	Swamp wallaby	
M. eugenii	Tammar wallaby	
M. parma	Parma wallaby	
M. rufogriseus	Red-necked wallaby	
Mirounga leonina	Elephant seal	
Mus musculus	House mouse	
Mustela ermina	Stoat	
M. nivalis	Weasel	
M. putorius	Ferret	
Mystacina tuberculata	Short-tailed bat	Pekapeka
Odocoileus virginianus	Whitetail	
Oryctolagus cuniculus	European rabbit	
Petrogale penicillata	Rock wallaby	
Rattus exulans	Polynesian rat	Kiore
R. norvegicus	Norway rat	
R. rattus	Ship rat	
Rupicapra rupicapra	Chamois	
Sus scrofa	Wild pig	
Trichosurus vulpecula	Opossum	

BIBLIOGRAPHY

A.A. Book of the New Zealand Countryside. Lansdowne Press, Auckland, 1978.
CHILD, J. *New Zealand Insects.* Collins/Fontana, Auckland, 1974.
FORSTER, L.M. *Introduction to New Zealand Spiders.* Collins, Auckland, 1973.
FORSTER, R.R. & L.M. *Small Land Animals of New Zealand.* McIndoe, Dunedin, 1970.
JOHNS, J.H. & POOLE, A.L. *Wild Animals in New Zealand.* Reed, Wellington, 1970.
KNOX, R.(ed.). *New Zealand's Heritage.* Hamlyn, Auckland, 1971.
KNOX, R.(ed.). *New Zealand's Nature Heritage.* Hamlyn, Hong Kong, 1974-76.
McDOWALL, R.M. *New Zealand Freshwater Fishes: A Guide and Natural History.* Heinemann, Auckland, 1978.
McLINTOCK, A.H. (ed.). *An Encyclopaedia of New Zealand.* Government Printer, Wellington, 1966.
MILLER, D. *Common Insects in New Zealand.* Reed, Wellington, 1971.
MORRISON, P. & HARRIS, L.H. *Forest Wildlife.* N.Z. Wildlife Service & N.Z. Forest Service, Wellington, 1974.
MORTON, J.E. *The New Zealand Seashore.* Collins, Auckland, 1968.
NEW ZEALAND WILDLIFE SERVICE. *Wildlife – a Review.* Wellington, 1970.
OLIVER, W.R.B. *New Zealand Birds.* Reed, Wellington, 1974.
POWELL, A.W.B. *Native Animals of New Zealand.* 2nd edn. Unity Press, Auckland, 1951.
REGENSTEIN, L. *The Politics of Extinction.* Macmillan, New York, 1975.
ROBB, J. *New Zealand Amphibians and Reptiles in Colour.* Collins, Auckland, 1980.
ROBERTS, G.S. *Game Animals in New Zealand.* Reed, Wellington, 1968.
ROYAL FOREST AND BIRD PROTECTION SOCIETY OF NEW ZEALAND. *Forest and Bird.* (Journal) Wellington.
SHARELL, R. *The Tuatara, Lizards and Frogs of New Zealand.* 2nd edn. Collins, Auckland, 1975.
STEVENS, G.R. *New Zealand Adrift.* Reed, Wellington, 1980.
THOMPSON, R. (ed.). *Bibliography of Offshore Islands.* N.Z. Oceanographic Institute, DSIR, Wellington, 1977.
TURBOTT, E.G. (ed.). *Buller's Birds of New Zealand.* 3rd edn. Whitcombe & Tombs, Christchurch, 1967.
WARDS, I. *New Zealand Atlas.* Lands and Survey Department, Wellington, 1976.
WILLIAMS, G.R. (ed.). *The Natural History of New Zealand.* Reed, Wellington, 1973.

INDEX

Figures in bold refer to illustrations